# The Global Energy Trap and a Way Out

**Frank Parkinson**

Ω

OMEGA
POINT PRESS

Originally published by Matador Books, 2017

Republished in the UK by
Omega Point Press
Lytham St Annes | Ulverston

ISBN 978-1-901482-03-4

In memory of Frank Parkinson
1928-2018

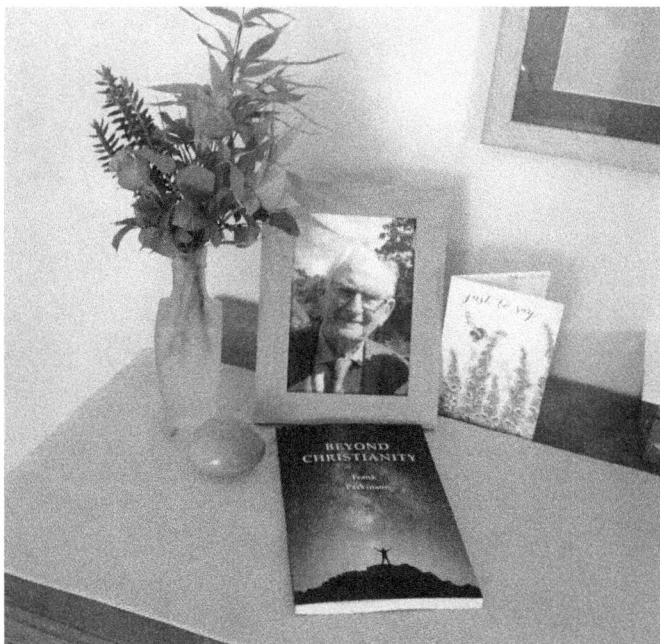

# Global Energy Trap

# Contents

Foreword                                                                  xi
Introduction                                                               1


**Part One: The Global Energy Trap**

**1 Energy and Civilization**
1.1    The Ongoing Energy Revolution                            13
1.2    The Importance of Energy Density                         19
1.3    Towards a New Energy Management Paradigm                 22

**2 Peak Oil and a Choice of Disasters**
2.1    Peak Oil: The End of an Era                              31
2.2    Global Warming and Climate Change                        36
2.3    The Mathematics of Disaster                              40
2.4    Global Warming and Big Business                          45

**3 Energy, Politics and War**
3.1    Oil and War                                             51
3.2    The Great Game and World War III                        56
3.3    The Game Changer                                        59

**4 The UK Crisis**
4.1    When the Lights go out in London                        63
4.2    The Political Response                                  65

**5 The Great Wind Farm Non-solution**
5.1    Costs, financial and environmental                      69
5.2    Finding the Truth                                       74
5.3    The Promise of Wind Power                               77

## Part Two: Fundamental Principles

### 6 Theoretical Foundations
| | | |
|---|---|---|
| 6.1 | E-plus in Concept | 85 |
| 6.2 | K-gen: A Metasystem | 89 |
| 6.21 | Collection | 89 |
| 6.22 | Conversion | 93 |
| 6.23 | Concentration | 94 |
| 6.24 | Conservation | 95 |
| 6.25 | Control | 96 |
| 6.3 | The Very Smart Grid | 97 |

### 7 K-gen and the Philosophy of Engineering
| | | |
|---|---|---|
| 7.1 | Engineering as a Social Force | 101 |
| 7.2 | K-gen and Systems Thinking | 102 |
| 7.3 | Learning from Nature | 105 |
| 7.4 | Three Principles of Energy Management | 107 |

### 8 The Economic Dimension
| | | |
|---|---|---|
| 8.1 | The Bigger Picture | 111 |
| 8.2 | The Economics of Domestic Energy | 117 |
| 8.21 | Solar Thermal Panels and Heat Pumps | 120 |

## Part Three: The K-gen System

### 9 The K-gen System: Parts and Functions
| | | |
|---|---|---|
| 9.1 | Going to Proof of Concept | 127 |
| 9.2 | Energy Collection | 127 |
| 9.21 | Solar Energy | 127 |
| 9.22 | Wind Energy | 129 |
| 9.23 | The Heat Pump Principle | 135 |
| 9.3 | Energy Concentration | 139 |
| 9.4 | Energy Conversion | 140 |

| | | |
|---|---|---|
| 9.5 | Energy Conservation | 143 |
| 9.51 | Insulation | 144 |
| 9.52 | Energy Storage | 150 |
| 9.53 | Waste Heat Retrieval | 156 |
| 9.54 | Voltage Optimization | 156 |
| 9.6 | Control | 157 |
| 9.7 | Outline of the K-gen System | 158 |
| 9.8 | Questions of Scale | 160 |
| 9.9 | Putting in the Numbers | 162 |

## Part Four: The E-plus System

### 10 The E-plus System

| | | |
|---|---|---|
| 10.1 | E-plus and the Evolution of the House | 167 |
| 10.2 | E-plus and a Cultural Shift | 169 |
| 10.3 | The Passive Energy House | 172 |
| 10.4 | The House as Energy Collector | 179 |
| 10.41 | Solar Energy Collection | 179 |
| 10.42 | Wind Energy Collection | 180 |
| 10.5 | Managing Light | 180 |
| 10.51 | Applying the Atrium Principle | 182 |
| 10.52 | Amplifying Through Reflection | 183 |
| 10.53 | Augmenting Window Apertures | 184 |
| 10.54 | Using LED Lighting | 185 |
| 10.55 | Exploiting Translucent Aerogels | 187 |
| 10.6 | Energy Conservation | 188 |
| 10.7 | New Approaches to Central Heating | 193 |
| 10.8 | The Retrofit Challenge | 195 |
| 10.9 | Harvesting Water | 196 |

## Part Five: The Scale of Things

### 11 An Engineering Revolution
   11.1   The Self-powered Engine     207
   11.2   A New Concept of Engine     208
   11.3   K-gen and the Anthropocene Era     210
   11.4   The RTE Principle and its Application     212
   11.5   The K-gen Prototype: A Heuristic Challenge     214

### 12 Raising Awareness
   12.1   Six Horseman     217
   12.2   Thinking Globally     219
   12.3   Conclusion: Escaping the Trap     222

## Appendices

**I Biochar and Chemical Sinkage: The Third Strategy**     227
**II Reducing Pollution from Transport**     237
**III Thorium: Safe Nuclear for the Future**     247

## List of Diagrams

Fig. 1    Energy loss in macrogeneration                              26
Fig. 2    Fossil fuel energy flow (steam engine)                     104
Fig. 3    K-gen renewable energy flow                                105
Fig. 4    Diffuser augmented cowled wind turbine (DAWT) 131
Fig. 5    Tornado wind turbine system (TWECS)                        133
Fig. 6    State change and latent energy                             137
Fig. 7    Closed loop state-change energy generation                138
Fig. 8    Solar dish Stirling generator                             141
Fig. 9    Simplified schematic of the Organic Rankine Cycle 143
Fig. 10   Approximate heat loss from average home                    144
Fig. 11   Flywheel storage subsystem                                 155
Fig. 12   Schematic for a K-gen system                               159
Fig. 13   Socrates "sun-tempered" house plan                        175
Fig. 14   Simple atrium schematic                                    183
Fig. 15   Angled window reveals                                      185
Fig. 16   Simplified schema of K-gen system                          208

# Foreword

## Professor Peter Wadhams, ScD.
Professor of Ocean Physics, University of Cambridge

When I first went to the Arctic, in the summer of 1970 aboard the Canadian oceanographic ship "Hudson", we were attempting a transit of the Northwest Passage. All along the north coasts of Alaska and the Northwest Territories, the Arctic Ocean sea ice lay close in to the land, leaving us a slot only a few miles wide to carry out our surveys. Sometimes the ice pushed right up to the coast and we had to break our way through, and eventually, when we were in the middle of the Northwest Passage, we had to be rescued by a government heavy icebreaker. In those days, a battle with sea ice in the Canadian Arctic was considered normal, even in high summer.

Today a ship entering the Arctic from the Bering Strait in summer finds an ocean of open water in front of her. The blue water extends far to the north, stopping not far short of the North Pole. The Northwest Passage is now easily navigable, and the year 2013 saw a record number of 203 ships sailing through it. In the 1970's sea ice covered 8 million square kilometres of the Arctic Ocean's surface; in September 2012 only 3.4 million. It is difficult to overstate what this means. Our planet has changed colour. We all remember the first beautiful photograph of planet Earth rising from behind the Moon, taken by the Apollo astronauts, a delicate blue sphere, isolated in the cosmos, which contains all that we know of life. That sphere was white at both ends. Today, from space, the top of the world in the northern summer looks blue instead of white. We have created an ocean where there was once an ice sheet. It is Man's first major achievement in reshaping the face of his planet, unintended and with dubious and possibly catastrophic consequences to follow.

If we look closer at the surviving ice, we see even more changes. The ice which is still left is thin: sonar measurements show that its average thickness dropped 43 percent between 1976 and 1999 and the reduction continues and accelerates. In the past most of the ice in the Arctic was several years old, and was called multi-year ice. It had a rugged and magnificent topography, with huge pressure ridges which blocked the paths of explorers and ships and which had keels reaching down 50 metres or more into the ocean. Today, by contrast, most of the ice is first-year; it has grown during the current season, reaching a maximum thickness of only 1.5 metres and with only a few shallow ridges to break up the very flat ice surface. Ice which grows during a single winter can now melt away during a single summer. It will not be long before the summer melt, driven by warmer air temperatures, outstrips the winter growth everywhere in the Arctic, and when that happens the entire remaining ice cover will collapse. We will have entered what the US climatologist Mark Serreze calls the "Arctic death spiral". It will leave us with an ice-free summer Arctic, though there is some disagreement about when this will have come about; computer modellers say in 20-40 years' time, perhaps in 2040, while real ice scientists who actually go out and do measurements are far more pessimistic and calculate that it will happen in less than 5 years, before 2020.

Whenever it happens, the consequences of a collapse of Arctic ice will be dramatic. Two huge effects are unleashed. Firstly, once the ice yields to open water, the albedo – the fraction of solar radiation that is reflected back into space – will drop from 0.6 to 0.1, accelerating warming of the Arctic and of the whole planet. Secondly, removal of the ice cover will take away a vital air conditioning system for the Arctic. So long as some ice is present in summer, however thin, the sea surface temperature cannot rise above 0ºC, since if the water warmed further it would lose its heat in melting some of the overlying ice. With the ice gone completely, the surface water can warm up by several degrees in summer by absorbing solar radiation, and over the shallow continental shelves this heat extends down to the seabed. This melts the offshore permafrost, frozen sediments which have lain there undisturbed since the last Ice Age. Thawing of offshore permafrost will act like releasing the lid of a pressure cooker; it will trigger the release of huge plumes of methane from the disintegration of methane hydrates trapped within the underlying sediment, and this will be a critical new development for methane has a greenhouse warming effect 23 times greater per molecule than carbon dioxide. It is indeed already happening, for an annual Russian-US expedition to the East Siberian Sea has observed methane plumes welling up from the seabed, and their observation has now been duplicated by Swedish and Norwegian work in the Laptev and Kara Seas. If this offshore methane release causes general atmospheric levels of the gas to rise, it will give an immediate and dramatic boost to global warming, for the area involved is huge, since one third of the Arctic Ocean is composed of shallow water shelves only 50-100 metres deep.

I have spent my entire scientific life from the age of 21 working on the science of sea ice and the polar oceans. What do these changes mean to me as I say a personal farewell to this magic landscape? Overwhelmingly I feel that this is a spiritual impoverishment of the Earth as well as a practical catastrophe for mankind. Our own greed and stupidity have taken away the beautiful world of Arctic Ocean sea ice which once protected us from the impacts of climatic extremes. Now urgent action is needed if we are to save ourselves from the consequences.

If we take a closer look at the consequences that will flow from this heating of the Arctic and the loss of its ice, firstly we find that it is warming 2-3 times faster than any other part of the world. We don't fully understand why, but it means that the immediate effect of enhanced Arctic warming is to cause sea ice retreat. Although some of the heating comes from the ocean, most is warming from the atmosphere. Both together have reduced the average thickness of Arctic ice by more than 40% between the 1970s and 2000, and if we combine this with the loss from its retreat we find that the volume of sea ice at the end of summer is now only 25% of what it was in 1979 when our satellite measurements began. From these data we can conclude that that the Arctic is driving global climate change as much as responding to it, and this makes it a particularly important factor in the now well recognized problem of global warming.

We can identify seven distinct global changes and feedbacks which stem from the loss of Arctic sea ice:

- A decrease in global average albedo, speeding up global warming
- An accelerated melt from the Greenland ice sheet leading to an enhanced rate of global sea level rise
- A snowline retreat in the northern hemisphere, further decreasing albedo
- A vegetation feedback which enhances warming through water vapor impact
- The serious threat of a major offshore Arctic methane outbreak
- The generation of extreme weather patterns in mid-northern latitudes, leading to a threat to global food production
- A decline in strength of the Atlantic thermohaline circulation.

We could call these the "seven global plagues" arising from Arctic sea ice loss. Each one is a positive feedback loop, where a change in sea ice extent initiates another undesirable or unexpected change in another part of the planetary system.

The albedo feedback occurs because as the summer ice retreats it opens up large areas of open water at a time of year when plentiful radiation is being received from the sun. The albedo of open water of is 0.1, much less than the 0.5-0.7 for melting ice, and this relative loss of sea ice between the 1970s and 2012 has caused a global albedo decrease, and consequent global warming equivalent to a quarter of all the carbon dioxide added to the atmosphere by man during that period.

The increased rate of melt of the Greenland ice sheet is thought to be due to warmer air masses moving over Greenland in summer, an added factor is now recognized. In summer pools of melt water began to appear on the Greenland ice sheet, with much of the water disappearing down holes called moulins which bring it down to deeper levels or to bedrock. At the same time the outlet glaciers began to accelerate, probably lubricated by this meltwater, so that some of them are now moving twice as fast, depositing much more ice into the ocean as icebergs. In the record year of 2012 there was a time in July when 97% of the Greenland ice sheet surface was covered with melt water, and a gravity satellite called GRACE, which measures the mass of the ice sheet, is finding that 300 cubic kilometres of ice are now being lost per year. This is a massive amount and means that Greenland is now the biggest source of water for sea level rise. Most glaciologists estimate that global sea level rise will be about one metre or more this century, possibly much more, and the process will be irreversible. It will have a disastrous effect on coastal cities like Miami, New York, Shanghai and of course Venice, as well as low-lying crowded coastlines like Bangladesh. Closer to home, there is every reason to think that the city of Hull will have become uninhabitable, as well as many smaller population centres on the coast.

As well as the climatic effect of retreating ice and snow cover, there is an important but less measurable effect from changes to the planet's vegetation cover, arising from the fact that high-latitude land masses covered with bare ground or low tundra plants are increasingly acquiring a more lush vegetation. Evaporation and transpiration from the leaves of these plants and bushes increases the level of water vapour in the atmosphere, and this adds to the warming effect since water vapour itself is a greenhouse gas.

However, the greatest threat to mankind comes probably through seabed methane emissions, a factor that is, surprising, almost completely ignored by the Intergovernmental Panel on Climate Change. It was, however, mentioned by the Pope in his recent encyclical *Laudato Si'*. We need to be especially worried about methane because, despite its much lower concentration in the atmosphere than the carbon dioxide that comes from burning fossil fuels, it has a much more powerful greenhouse effect. A sudden release of a large quantity of methane would have a huge, if short-lived, impact on the climate and the leaders of the Russian-US expeditions to the East Siberian Sea, Natalia Shakhova and Igor Semiletov, estimate that 50 gigatons (equal to $10^{12}$ tons) of methane could be emitted from the East Siberian Shelf during the next few years. This is, in fact, a conservative estimate based on their conclusion that the total volume of methane trapped in the sediments amounts to 720 Gt and is predicted to be emitted over ten years.

What this would mean in terms of global warming and economic cost to the world has been calculated by two colleagues and myself, our paper, in *Nature*, concluding that an extra warming of 0.6°C would occur by 2040, on top of the rest of global warming. Along with this, an economic analysis by my co-authors, using a model employed by the British Government, estimated a vast cost to the world of 60 trillion dollars over a century. This alone would have a disastrous effect on our attempts to limit the rate of warming of our planet, but because methane release on this scale is a new phenomenon, and has probably never occurred since before the last Ice Age, there are still many geophysical and climate scientists who still deny that it is possible.

Another huge threat to our planetary wellbeing is the likelihood that Arctic warming and sea ice retreat have been the cause of extreme weather patterns which have occurred over the past three years, typically involving very cold or stormy weather in winter in certain parts of Europe and North America and very warm weather in others. The jet stream, which is the fast-moving boundary flow separating Arctic from lower-latitude air masses seems to have slowed, due to the narrowing temperature difference between the tropical latitudes and the fast-warming Arctic. As the flow gets weaker the jet stream follows a wavier path, bringing cold air masses down to lower latitudes in the southward lobes and warm air to unusually high latitudes in the northward lobes. The slow movement of these lobes enables prolonged persistence of a local weather system in one mode, e.g. drought, flooding, cold weather or heatwaves. The biggest effects are occurring in mid-northern temperate zones, which is exactly the location of the planet's most productive croplands. If the effects persist, this could be a serious threat to global food production, with both direct consequences in the form of famine, and indirect in the form of social unrest in poor countries due to rises in food prices.

The seventh and final "global plague" is the thermohaline circulation, or "global conveyor belt", which is a very slow circulation of the world ocean. It is driven not by winds but by the differentials in distribution of heat and precipitation over the oceans and is at work continually worldwide. It has been called a conveyor belt because there are surface and deep components, with areas of upwelling or sinking to connect them, like the driving cogwheels in a conveyor. In the Atlantic the surface current comprises a small addition to the Gulf Stream, taking water from the tropics northeastward to bathe the coastline of Europe and then head further north still. Up in the Greenland Sea some of it sinks in a very small region at 75°N 0°W, and this is one of the main cogwheels of the conveyor. But it is a cogwheel that is failing, since the sinking had been driven by ice formation in winter, with the surface water gaining extra density from the salt that is left in the water when ice forms. In this special region the extra salt was just enough to drive deep cylinders of sinking water called "chimneys". But since 1998 no sea ice has formed in this region, and we suspect (though nobody has funded us to go there and find out) that chimneys can no longer form and so the conveyor belt is failing. On the assumption that it is indeed becoming weaker, the European Environment Agency estimated that by the end of the century Britain, Ireland, Iceland and the French and Norwegian coastlines will experience "only" 2°C of warming, compared to a ruinous 4°C for most of continental Europe, the difference being the loss of heat from the conveyor belt. This is in part good news for the UK, but not for tropical America, since the loss of this current will increase the temperature of tropical Atlantic waters, and hence increase the intensity of hurricanes.

The global feedbacks created by Arctic ice retreat are of enormous importance and public awareness needs to be raised of how they are affecting the earth's climate, agriculture, health and human existence in a profound way. Despite the growing evidence for potentially devastating consequences, there are still those who argue that the retreat of sea ice is of economic benefit because it opens up new routes to marine transport and helps with oil exploration in Arctic regions. The reality, however, is that we are living in a fools' paradise if we imagine that future climate warming can be modelled only in a linear way on carbon dioxide emissions. The reality is that new feedbacks come into play at certain critical stages, which accelerate warming and may end up dominating the future pattern of global change. We must be aware of the two emerging feedbacks which pose dangers as great as the increase in atmospheric carbon dioxide, namely, loss of albedo feedback and impending methane emissions. Albedo feedback is real and definite and is increasing global warming by 50%. Methane feedback is postulated for the near future and will, if it happens, more than double the warming rate. So it may be that even if we radically reduce carbon dioxide emissions, the system may not respond – it is developing a momentum of its own.

It is a cause for the greatest concern, from both a moral and practical standpoint, that these effects are being downplayed by the very body, IPCC, which was established to warn the world of dangerous climatic change. Its view is that if we reduce our carbon emissions at a rapid rate we can save the world from dangerous climate warming. I wish that I could agree with this view but my own conclusion, based on Arctic feedbacks, is that even a rapid reduction in $CO_2$ emissions will not work in time, so we must seriously and urgently consider emergency methods which could slow down the rate of warming and give us time to change to a new paradigm of living on this planet. This means applying geoengineering techniques, unattractive as these may be to many people including scientists. Geoengineering can be thought of as a sticking plaster solution to our global problem. It consists of reducing the radiation absorbed by the planet, typically by spreading finely divided powder in the stratosphere to reflect incoming solar radiation or, more benignly, by injecting fine water droplets into the lower parts of marine stratus clouds to make them brighter, that is, increase their albedo. In the last analysis it is not a permanent solution. It does not actually halt the growth of $CO_2$ in the atmosphere, so as soon as the treatment is stopped the disease (rapid warming) breaks out with even greater virulence. It also does nothing to halt the acidification of the ocean, another product of increased $CO_2$ levels, which will wipe out coral reefs and seriously impact marine life.

The only actual *solution* to catastrophic global warming, apart from an impossible appeal to mankind to cease emitting $CO_2$ altogether and immediately, is to find a way to *take CO2 out of the atmosphere.* This is the ultimate technofix, but those wise souls who advocate that we live in closer harmony with nature find the concept difficult to accept, because it would allow us to continue with all our bad habits such as continued fossil fuel use. Various methods, feasible and not so feasible, some of which are reviewed in the appendix to this work. Among them are the systematic use of biochar, which the book recommends, massive worldwide tree-planting, carbon capture and storage from coal-fired power plant exhausts and even the exposure to the atmosphere of untold billions of tons of crushed olivine rock, a common rock type which very slowly undergoes a chemical reaction in air which involves absorption of $CO_2$. A straightforward cost-effective and energy-effective method has not yet been invented, and this is a true challenge to our human ingenuity. Can we persuade our politicians and scientists to have a new Manhattan Project, a focus for a massive worldwide research effort, to design an effective method of removing $CO_2$ from the normal atmosphere and turning it into a benign substance which can be stored or used? To my mind this is the most important challenge in science and technology today, since our very existence is at stake. We have created global warming and we ought to be able to stop it.

This book offers a plausible way to escape the fatal straitjacket of accelerating climate change, by making our own homes into sources of emission-free energy, so that we are not passively involved in making the problem worse. It is by no means the only work on this theme, which is rapidly emerging as a social concern, but it stands out because it envisages a world where this is a norm, and where the ordinary house, and larger installations, produce baseload electricity from renewable energy sources. If, when all the research and development has been done, this proves possible, we shall have a realistic chance of bringing climate change under control.

Cambridge, August, 2016

# Introduction

## Energy Challenge and Response

*There is no substitute for energy. The whole edifice of modern society is built upon it .... It is not 'just another commodity', but the precondition of all commodities, a basic factor equal with air, water and earth.*

E. F. Schumacher [1]

*We face a crisis. If we are to rebalance our energy needs with what we can produce cleanly through renewable sources, we need a revolution in terms of the way we build.*

Baron Norman Foster [2]

It is said that the Teflon coated frying pan was a spin-off from the NASA moon rocket programme. This book is more in the nature of a moon rocket spun off from a frying pan project. It began as little more than a checklist of items which could improve the energy-efficiency of the ordinary house, bringing together various familiar items such as solar panels, heat pumps, triple-glazing and energy-saving light bulbs. What seemed at first a simple and well-defined task started to change shape as researching the literature revealed the number and variety of these items. To add to the difficulty, it soon became clear that this was a moving front in many respects, for improvement is ongoing and economies continually improving. A solar panel, for instance, that cost £4,000 in 2008 costs little more than half of that sum in 2015 and is fifty per cent more efficient. Then, as the range of energy generating devices, not just solar panels, came into view, minimization of energy consumption and maximization of clean energy generation became separate challenges. The first was essentially architectural, the second called for engineering of an imaginative, indeed pioneering, nature.

It soon emerged that the engineering challenge was of a quite different order of complexity from the architectural improvements and was, in fact, leading them, as so often in the history of architecture. Then it became obvious that the engineering component, initially conceived as an integral part of the house could have a much wider application if it were approached as freestanding and scaled up. So it was decided to split the original project and give the systems different names. The architectural challenge was identified and defined by naming it the *E-plus System*, which would allow it to be treated as part of the normal toolkit of a qualified architect, much as the Passivhaus recommendations have come to be used as a reference book. The engineering challenge, which was essentially how to manipulate and store energy collected from renewable sources was named the *K-gen system*. The K stands for "kathar", which is the Greek word for "clean".

The global energy trap largely concerns the build-up of carbon dioxide in the atmosphere, and as the implications of the carbon cycle were studied, a third component took on more importance. This was the potential, and the necessity, of carbon sequestration to counter global warming. A coordinated initiative to convert biomass into biochar (essentially charcoal) and bury it, if carried out on a large enough scale, could have a dramatic effect on rebalancing the carbon cycle. However, promising as such a strategy undoubtedly is in reducing the atmospheric overload of carbon dioxide, it was decided that to give it a proper treatment would weaken the focus of the book. Hence, having drawn attention to it in this introduction, the biochar revolution now in the making has been relegated to an appendix. This may prove to be ironic, since it could well be at least as important in escaping the global energy trap as the two themes on which the book concentrates.

Although a substantial part of the book is devoted to explaining the energy disaster which the world now faces, this is an optimistic work. The E-plus and K-gen systems together constitute a novel approach to energy generation, transformation and storage, and while biochar is the solution to a different kind of problem, all taken together constitute a three-pronged plan of action to combat global warming and fossil fuel energy depletion. Al Gore has said that there is no "silver bullet" answer to the complex problem of energy depletion and global warming, but the book is offering three silver bullets that can do the job. To determine the plausibility of this claim, one needs to begin with an understanding of what the global problem actually is and why it is described in the title as a trap. That task constitutes Part One of the book. It is actually multiple crises which interlock and are converging towards a point of catastrophe between thirty and forty years from now. Unless massive action is taken before then, we, or our children, may expect to experience global breakdown on a scale which may truly be called apocalyptic. That is a sensational term, but if it actually describes future events, there is nothing to be gained by seeking comfort in a less emotive word.

The overarching theme is that burning coal, oil and gas at current rates is leading to atmospheric pollution, which is the main cause of global warming, and that in turn is leading to climate change, the effects of which we are already experiencing. If not checked before it reaches a tipping point, when Antarctic ice a mile deep begins to melt, future generations will live on a planet where the oceans will have risen not by a few centimetres but quite possibly by several metres. If this should happen - and that is what current projections are pointing to - vast areas of low-lying land, including many great port cities, will be deep under water and, hard as it may be to imagine, most of London will have become a latter day Atlantis. The data-based prediction given by Professor Wadhams in the foreword is backed by many studies, a 2015 report by the International Panel on Climate Change (IPCC) suggesting that in the long term for every one degree rise in global temperature we may expect a rise of 2.3 metres in the ocean level.[3] Whole countries could disappear from the map, as, for instance, Bangladesh, 75% of which is less than ten metres above sea level. Flood barriers or embankments could not cope with this sort of happening, incremental though it may be, and an exceptional storm surge is likely to make it happen sooner rather than later. The higher habitable ground of the globe, now greatly reduced in area, will have to accommodate not only those who have been displaced but, if present demographic projections are right, four billion more souls by the end of the present century and on decreasingly fertile soil, much of it turning to desert. Climate trends already evident will intensify, and formerly temperate climates will be ravaged by frequent and prolonged heat waves, droughts, exceptional downpours of rain and flooding, more frequent and more destructive hurricanes and the failure of age-old monsoon patterns. There is good reason to think that the Gulf Stream, which gives western Europe mild winters and summers, will cease to flow and a city like Edinburgh will experience a similar kind of climate to Moscow, which is on the same northern latitude. Professor Wadham's foreword gives some initial evidence for these developments, which many might otherwise find unbelievable and dismiss as science fiction or scaremongering, and he makes the very important point that if such future developments are within the bounds of reasonable probability, it would be irrational, and deeply irresponsible, not to take action now.

By a painful irony, while all this is happening, the fossil fuels which have brought it about, having done their damage will be depleting to exhaustion, and unless replacement sources of energy are found, there will be equally devastating social consequences. This is the global energy trap of the title. A civilization which has been built on cheap and seemingly unlimited energy from coal and oil will have to reshape and reconstruct itself in ways as yet unimagined. The very foundations of economics and society will have to be re-laid, for economic activity is of its nature an exchange of surplus production and the prosperity of the world today comes from a surplus which is entirely dependent upon the serendipitous discovery of fossil energy sources and the invention of the external and internal combustion engines that put this energy to work. Greek and Roman civilization worked because they had the "free" energy of slaves, over a third of Athenians at one time being in this class. Today, as is often noted in economic textbooks, we each have the equivalent of a hundred slaves at our disposal, thanks to coal, oil and gas.

Britain finds itself in a rather unique position in the perfect storm of energy-related crises that we can now see approaching, for as privatization has broken up the old nationally owned utilities, the nation has increasingly found itself without a coherent energy policy. The consequences of this lack of government direction are now apparent in the fact that fully a third of present electricity generation will be taken out between 2012 and 2022, with no replacement planned, other than nuclear installations that are unlikely to come on stream before 2024 and the promise of a bonanza of gas from the fracking programme pushed forward by the government in the face of very serious dangers of long-term pollution of the water table. In a quixotic burst of activity the government has committed itself to a programme of building wind farms, a decision so monumentally misguided that it calls for closer analysis, which is given in Chapter 5, *The Great Wind Farm Non-solution*. What is notable in all the naive enthusiasm for wind farms is a near-total lack of research into storage of the electricity so produced, which does not arrive when needed and which wind farm owners are often paid to dump. The K-gen system will say enough about new ways of storing power to justify initiatives on the large scale which the book will introduce on the small scale.

Until the E-plus and K-gen systems are designed, tested and applied on the small and large scale, judgement on their claims must obviously be suspended, but those claims alone set it apart. The energy-saving E-plus house, as conceived, incorporating the K-gen electricity generating system, is designed not only to provide all the energy that the average house needs, but a surplus which will be fed to the National Grid, when required and in the amounts required, unlike renewable energy initiatives in general, which produce electricity only when the sun shines or the wind blows. If half a million homes could be so designed or adapted to do this, the impact of the E-plus house on the nation's energy supply would be too great to ignore, but if that figure were to be raised at some future time to ten million homes, the shape and function of the National Grid would be irrevocably changed.

The big question then is, *Can it be done?* It is hardly an exaggeration to say that the challenge of converting ten million dwellings to E-plus standard in the UK alone, building perhaps thousands of stand-alone K-gen systems and at the same time organizing a biochar programme would be literally at the level of mobilising for war.

The need for action on this scale in the face of Britain's energy crisis has, in fact, been made in exactly these terms by Jeremy Leggett, a prominent crusader for clean energy and chief executive of Solar Century, which manufactures photovoltaic equipment:

> *We need an emergency war cabinet appointing an army to help people conserve energy, invest in solar and all other sensible alternatives. We've got to build the new Spitfire.*[4]

The same sort of imagination that resulted in the development of radar, the long range ballistic rocket and Mulberry Harbour in World War II needs now to be stimulated and applied to harnessing renewable energy and developing micro- and mesogeneration as a source of baseload electricity. The existence of a National Grid acts silently to convince even pioneers in this field that top-down generation is natural and necessary and while microgeneration is widely recognized as worthwhile, it is generally assumed to be an add-on, at the margin of the greater energy picture. This book is presented as a first step to putting it firmly at the centre.

The push to microgeneration has important implications for society, in general reinforcing and extending E.F. (Fritz) Schumacher's crusade for a downsizing of industrial and commercial units, as argued in his most famous work *Small is Beautiful*. It also draws upon his less well known writings on energy management and goes beyond him in advocating a "Small is essential" approach to electrical generation, not only to supply national requirements but to counter the crisis of climatic change. It is fair to say that in the period during which he was writing on energy matters, from 1950 to his death in 1977, neither Schumacher nor anyone of note foresaw the problem of atmospheric pollution now facing the planet. This is understandable, since in his day car-ownership was limited - perhaps five cars per hundred people in the UK, as against forty today - and the Clean Air Act of 1956 had effectively done away with the dreadful atmospheric pollution problem of the day, namely, airborne soot from coal burning that fouled industrial towns and cities. The great London "smog" of 1952 was so bad that it closed down city traffic for four whole days and killed many thousands of people and finally moved the government to action. We will probably need something equally traumatic before the UK and other governments recognize the national importance of the global pollution that is now building up inexorably.

Schumacher was farsighted in his awareness that energy played so great a part in national wealth and wellbeing that it could not be left solely to market forces, and he pressed the case for public corporations, of which the National Coal Board was an early example and something of a test case. He was, however, equally aware of the sclerosis that too easily sets in when large scale bureaucracies are created and he thought "out of the box" about this endemic problem, coming to the conclusion that public corporations should be so designed as to make a profit. Although this seems at first contradictory, it is, in fact, a win-win solution, justified by the fact that profit made from the community is in effect a reduction in general taxation.

It could, of course, be said that as Chief Economic Adviser to the National Coal Board Schumacher was biased in seeing the need for government oversight of energy policy, but as the need for a coherent and large scale strategy for renewable energy becomes apparent, his arguments carry even more weight now than then. After thirty years of full bore privatization, the government finds itself having to go into reverse and try to reconnect the different bits of the energy sector in order to have a national energy policy that works. Schumacher actually saw Peak Oil and its consequences before Marion King Hubbert invented the term, but at the time when he was writing, it was too far away to have practical consequences. In a 1958 paper to the Royal Statistical Society he dealt with the long term view of energy management, using the phrase "the twilight of the fuel gods", but oil and coal were too cheap and abundant at that time to warrant giving full attention to a policy of renewables, and apart from a few fringe architects, who will appear in the book, the subject was almost entirely neglected. That said, "energy crisis" was certainly on Schumacher's radar, and was, in fact, in the title of a paper delivered to the Royal Swedish Academy of Science as early as 1973. In it he talks of the undeveloped potential of "solar energy and its derivatives and also of tidal power and geothermic heat" and looks forward to "the development of 'new styles of technology'." [5]. Because Schumacher's name is associated with the phrase "small is beautiful", it is sometimes forgotten that his main emphasis was on "appropriate technology", and a technology that has been dedicated to utilizing the energy of the molecular bonds of fossil fuels and the subatomic bonds of uranium and thorium is no longer appropriate. What now is appropriate and urgent are advanced technologies to utilize the superabundant energy of sun, wind and material phase change and extract carbon permanently from the atmosphere. Everything about the man signals that he would throw his weight today behind the search for a coordinated policy of design and production in these areas. He should be living at this hour.

The book is divided into five parts, with the first part setting the social and economic scene for engineering and architectural issues to follow. Part one, *The Global Energy Trap*, deals broadly with the nature of the energy crises that the world and the UK now face and explains why the expression "global energy trap" is appropriate to describe a situation where the world economy will collapse if we run out of fossil fuel energy before finding a clean substitute and its atmosphere will be irreversibly wrecked if we use up all the remaining fossil fuels. Part two, *Fundamental Principles*, introduces the proposed solution and explains how it is best understood as an exercise in systems thinking. Parts three and four, *The K-gen System* and *the E-plus System*, identify the specific engineering and architectural challenges, puts forward ways in which they can be met. Part five, *The Scale of Things* consists of two chapters, recapitulating the main themes with selected emphases, to make clear that in the proposed strategies we have a realistic way to solve the planetary problems of energy depletion and, in the longer term, of atmospheric pollution. There are three important appendices, including the one on *The Biochar Revolution* mentioned above. *Reducing Pollution from Transport* deals briefly with alternative fuels, notably methanol and ammonia, and *Thorium: Safe Nuclear for the Future*, as its title suggests, deals with an alternative to plutonium-fuelled reactors. It should perhaps be emphasized that although the present book contains many technological proposals, its main purpose is to raise awareness as a prelude to action.

### References

1.  Geoffrey Kirk, ed., *Schumacher on Energy.* London: Jonathan Cape, 1982. p. 1.

2.  "Norman Foster on Building a Sustainable Future," *The Telegraph,* 5/9/2011

3.  Quoted in, Michael le Page, "Five metres and counting: Only drastic action will prevent a 20 metre rise," *New Scientist,* 13 June, 2015

4.  Jeremy Leggett, CEO Solar Century. Interview with Alex Smith of the Post Carbon Institute *Energy Bulletin.* 25/2/2011.

5.  Kirk, *Op.cit.* pp. 52-53

# Part One

# The Global
# Energy Trap

*The laws of thermodynamics
control, in the last resort, the rise
and fall of political systems, the
freedom or bondage of nations,
the movements of commerce and
industry, the origins of wealth
and poverty and the general
physical welfare of the race.*

Frederick Soddy,
***Matter and Energy*** (1912)

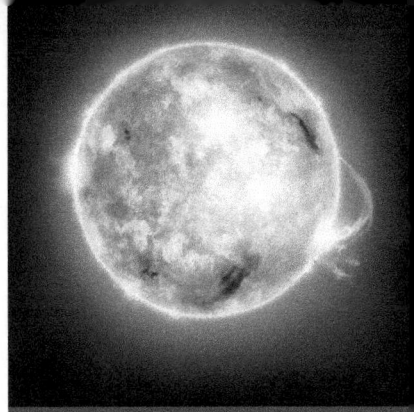

# Chapter 1

## Energy and Civilization

*We are in a crisis in the evolution of human society. It's unique to both humans and geologic history. It has never happened before and it can't possibly happen again. You can only use oil once. You can only use metals once. Soon all the oil is going to be burned and all the metals mined and scattered. [Our] window of opportunity is closing ... and the time scale is not centuries, it's decades.*

M. King Hubbert [1]

*Our freedoms, our comforts, our prosperity are all the products of fossil carbon, which is primarily responsible for global warming. Ours are the most fortunate generations that have ever lived [and] might also be the most fortunate generations that ever will. We inhabit the brief historical interlude between ecological constraint and ecological catastrophe.*

George Monbiot [2]

### 1.1 The Ongoing Energy Revolution

It could be said that the modern age began three hundred years ago almost to the day with the rough diagram of a steam piston engine made by Thomas Newcomen. From this unscaled and unquantified freehand sketch he went on to make working drawings and construct an engine in 1711 which was able to pump water from a coal mine several hundred feet underground. This new invention could replace hundreds of horses previously used for the same work. Fifty years later James Watt made critical improvements to Newcomen's model, which improved the fuel efficiency and decreased its size and, with the financial backing and marketing skill of Matthew Boulton, went on to manufacture or license hundreds of his engines and sell them worldwide. This gave critical impetus to the Industrial Revolution, which was to result in a new civilization. The Industrial Revolution was at base an energy revolution.

Now we are entering into a new energy revolution - and a future new civilization – as the planet's fossil fuels become exhausted and atmospheric pollution caused by their waste products threatens planetary disaster. The seeds are being planted for a new vision in engineering, focusing on the need to exploit clean, renewable energy to the full and a new understanding of the theoretical foundations of energy management. In the latter respect the present situation may be contrasted with the first energy revolution, which brought the steam engine into existence almost without theory. Indeed, it would not be untrue to say that the theory of energy and the science of thermodynamics emerged in large part as an attempt to understand what the engineers had already achieved. The unit of energy at this time was based on the number of horses that a steam pumping engine could replace, and the unit of horse power is still with us. This time, as the present work aims to show, it will be largely the other way round: the engineering will build on a coherent theory of energy. That is, however, to put it too simply, for there is always a creative synergy between theory, observation and experiment, in engineering as in science, and indeed perhaps in all human affairs.

The nature of energy itself has always been puzzling. Until the end of the nineteenth century there was no universally accepted theory of the atom, or even agreement that atoms existed, and consequently no awareness that the energy which drove the steam engine came from the release of the chemical bonds of coal. There was no formal systems theory at the time and, as a consequence, the laws of thermodynamics were put together *ad hoc*, biased towards closed system thinking and heavily influenced by the observation that useful fuel turned to useless ash and in temperate climates hot things tend to become cold. That is a common understanding of the concept of the law of entropy which governs physical science today, and about which no less an authority than Sir Arthur Eddington expressed the view that this law is not only unbreakable but unquestionable. In his words, "The law that entropy always increases - the second law of thermodynamics - holds, I think, the supreme position among the laws of Nature."[3] The theory of energy is still based on the obviousness of fuel entropy in much the same way that astronomy was for many years based on the obvious fact that the sun went round the earth. Had the pioneers of energy theory lived in a hot desert, they may have paused to consider that cold things obviously become hot, but the point at issue here is that energy management theory has its theoretical foundations in the economic realities of burning fossil fuel to generate useful energy. The most significant facts which it embodies are that fuel is spent, hot turns to cold, the system runs down, order turns to disorder, energy becomes too diffuse to perform work. All these things combine to create a mindset which now has to change if we are to harness for man's use the renewable energy of nature and the latent energies which are released in state change. The main reasons for this are two: firstly, renewable energies are generally diffuse and of low density and, secondly, there is a category of quasi-renewable energy in the form of the latent energy of state change, or phase change, the critical importance of which has not yet been recognized. Since less energy is required to trigger state change than the energy that is released, a system that exploits this fact can in effect provide something for nothing. The foundations of energy theory now need to be re-examined and in some respects re-laid to take account of these things.

The present work, though small, is revolutionary in another important aspect, namely, in its assumption that the very nature of human society, and therefore of the human individual, is determined by its access to energy and the way it is used. This is the point of the quotations from Marion King Hubbert and George Monbiot at the head of this section, which were very deliberately chosen to make the significance of this book as compactly as possible. It was Hubbert's meticulous analysis of the data which led him to conclude that oil, the prime source of the world's energy, would be exhausted within the lifetime of our children - and what then? Unless some new, abundant, cheap and reliable source of energy is found, we can expect at best a civilization in decline and at worst economic and social collapse. As Appendix 1, *Biochar* explains, there is every reason to think that a partial replacement may be available in bio-oil, but

the main point is that one day in the not too distant future petroleum will run out. Monbiot, of a later generation, emphasized the
ecological degradation that will necessarily follow if global warming
continues at its present rate. These are the crises that are converging
to catastrophe which will be described in more detail in the following chapter. They are not the only ones facing the human species, for
overpopulation is an even more intractable problem, and intensifies
the energy crisis, but that has been left out of the present treatment
to keep the line of argument as clear as possible.

It cannot be too strongly emphasized that insofar as every individual is shaped by society, the topic of energy is ultimately inseparable from what it means to be human. The proposition, which at
first may seem to be exaggerated, is seen more clearly if one envisages
how being born into a caveman culture will result in one thinking
and acting like a caveman. By the same line of reasoning, someone
born into an energy-rich society will be a different kind of person
to one who has to spend a large part of his or her life seeking a few
sticks to make a fire or relying on dried cow dung for fuel. Indeed, in
such cases there can hardly be anything worth calling culture, since
education and technology in any meaningful sense can never be developed. Following through this logic, one would expect a growing
availability of energy to result in cultural changes and opportunities,
and so it is seen to be on the most superficial examination of the history of energy revolutions.

The full significance of energy in relation to human affairs is
a historical saga, and a strong case can be made for arguing that
no one should leave high school without being introduced to this
remarkable story of how energy revolutions have shaped the modern world. The most significant development in energy use came
with the widespread replacement of wood by coal, which occurred
in the industrialized world around 1800, and was both the cause and
consequence of the industrial revolution. Supplies of fuel wood in
Europe were running dangerously low, as the population increased,
and in the UK the demand for timber to build an expanding im-
perial navy – the famous "wooden walls" – was putting a critical
pressure on supply. Coal was well known as a fuel, but not easy to

exploit, since water rapidly collected in underground workings and, until the invention of the steam engine, could rarely be pumped out. Once coal became widely available to drive the engines that drove the pumps which made the coal available, a virtuous feedback was put in place and the industrial revolution began in earnest. Coal is still very much with us, since coal-fired stations produce well over half of the world's electricity, and an even greater proportion of atmospheric pollution, but it began to be replaced by oil in the late 1800's, as oil's cleanness, its greater energy density and transportability became recognized.

With the use of petroleum products for fuel there came a far-reaching social change. As well as enabling a great leap forward in industrial efficiency, the impact of oil on the individual's life was to open up a new dimension of empowerment, and free up perhaps ninety per cent of the population to engage in cultural activities of one kind or another, be it creating universities or symphony orchestras or sending a man to the moon. To take only one example, a man with a tractor can farm fifty times more efficiently than one man with a horse plough or scythe. As I write these words, I can see through the window someone singlehandedly unloading five tons of paving blocks with a hydraulic hoist powered by the diesel engine of his truck. Within the space of an hour one person has been able to load the truck, drive it several miles and drop its load precisely where required, without raising a sweat. It would be hard to calculate how many back-breaking man hours this would have required a century ago - perhaps a thousand, perhaps two, not to mentions several horses and wagons. This and similar miracles are now completely part of our world-taken-for-granted, but few stop to wonder what will happen to society when the oil runs out. Until then we are criminally profligate in our use, and abuse, of it. We ship our beans from Kenya and our apples from Chile, regardless of the atmospheric pollution thus caused. George Monbiot cites the quite normal instance of supermarkets in Evesham selling vegetables grown within walking distance of the supermarket but arriving there only after being trucked to Herefordshire, then to a packing factory in Wales, then to a distribution centre in Manchester and finally back to Evesham - a total journey of some 670 kilometres. Only with cheap oil could we have come to create such a lunatic world.

Oil and its by-products, such as plastics and asphalt, have been the obvious driver for modern civilization in making transport for the masses possible; where once private transport was only for the rich and the nobility, any Tom, Dick and Harriet can have their "horseless carriage", in the West at least. Indeed, the motor car has become so cheap that a very usable second hand model can be purchased for a fortnight's wages. Add to that the way in which human identity is slowly being changed through globalization, in large part made possible through air transport, and one starts to appreciate the evolutionary impact of oil as fuel. Indeed, man's discovery of petroleum has literally changed the face of the earth and made him a geological force. The Suez and Panama canals, dug out with huge expenditure of coal energy have reshaped the planet's oceans, but other excavations on an even greater scale are now taken for granted. The Kalgoorlie Super Pit, where opencast mining for gold has been carried on for over 150 years, is 570 metres deep, well over half a mile, and measures 3.5 km x 1.5 km. That may at first seem a stupendous hole, but is only a quarter as large as the Chuquicamata copper mine in Chile, which is still not the world's largest. If the cavity in the earth made by the Bingham Canyon open pit mine in Utah were to be turned inside out, so to speak, and considered as a mountain, it would be almost as high as Mount Snowdon. All these, and other things like huge dams, are geological features observable from outer space, and all these modern geo-scale excavations have been made by releasing the energy contained in the molecular bonds of billions of tons of diesel oil.

On top of the increase in industrial productivity made possible by oil has come a doubling and redoubling of crop yields by the application of fertilizers, whose feedstock is natural gas, and of pesticides, which consume about 4 per cent of the world's oil, and one starts to understand how dependent modern society is on the combined advantages of oil- and gas-sourced energy. This has the unseen but entirely predictable consequence that organizations which make huge profits from the extraction, refining and sale of oil and gas, not to mention the countries which own the reserves, have a vested interest in the present situation continuing without change. They will oppose, as far as possible, development of alternative energy sources, which threaten their monopolistic profits, and ignore the damage that is caused to the planet and the health of its inhabitants from the polluting waste products of oil. Their opposition is well organized, well funded and chronicled.[4] What organization, then, should take the initiative? Who will speak for our planet? These questions are implicit in the theme of this book. The answer, in a word, lies not in government initiatives, for most departments of energy are in thrall to Big Oil and Big Coal, and have little long-term strategic vision, but in millions of "small individuals", who will solve the problem by investing in a renewable energy system for their own benefit, once such a system is available. That is the main economic thrust of the book. If this logic plays out in real life, we shall have a modern day version of Adam Smith's "invisible hand".

## 1.2   The Importance of Energy Density

The energy revolution that we have enjoyed for two centuries has been driven by two main factors, namely the abundance of the new fuel sources of coal and oil, which for a while seemed to be limitless, and the fact that each newly discovered fuel was of greater energy density than that which it displaced. When it comes to renewable sources of energy, there is unlimited abundance of sun and wind, but in its natural state, as it were, it is diffuse and without much practical use. To make it more usable is the most obvious engineering challenge, and that in turn will call for concentrating what might be called the energy content of these natural sources and converting it into electricity.

The table below shows the energy densities of different fuels for purpose of comparison. It is necessarily simplified, since there are big differences between, say, green and dry wood and bituminous coal and anthracite. Uranium stands out dramatically and the figure is likely to be even higher in the case of fast breeder reactors, but the nuclear catastrophe at Fukushima, plus the seemingly insoluble problem of nuclear waste disposal, suggest that the current building programme of nuclear power installations worldwide will slow and eventually end within perhaps the next thirty years. The nuclear revolution is a separate and controversial topic, but it is worth noting that its future may lie in using thorium, rather than uranium as a fuel, on which more will be said in Appendix III. At present, some countries, notably France, are able to supply almost all their electricity requirements, and produce an exportable surplus, from nuclear power, but, on the other hand, the problem of long term waste storage will not go away and tends to be swept under the carpet. Also, while it is argued that Fukushima is a "black swan" event, so improbable as to be effectively a one-off, the effects of global warming in creating increasingly extreme weather conditions is a real cause for worry. In September 2013 Colorado, to take but one instance, received 80 per cent of its annual rainfall in four days, with devastating effects, and the exceptionally high Missouri floods in spring 2011 came very near to putting the Fort Calhoun Nuclear Facility near Omaha, Nebraska at risk. Safety is far less of a problem with thorium, since the process embodies automatic switch-off in case of failure and the half life of spent fuel is far shorter than that of uranium.

The last three items in the list below - ethanol, methanol and ammonia - have been included because they will appear in Appendix II as complementary renewable fuels which suggest at least a partial solution to global pollution from fuels used for transport. However, the use of alcohol biofuels or ammonia is a separate subject too vast to be treated here in any depth and little more can be done than point the reader in the direction of source material. Hydroelectric power has been omitted, since it cannot be measured by this standard and is not of relevance to most countries in the world, which do not have either the geographical or climatic conditions to make use of hydro.

| Wood | = | 2.8 kilowatthours (kWh) per kg. |
| Peat | = | 3.5 |
| Coal | = | 9 |
| Petrol | = | 13 |
| Uranium | = | 830,000 |

Ethanol    =    6.8
Methanol   =    5.6
Ammonia    =    3.6

A word should be added here about units. Following the lead of David MacKay in *Sustainable Energy - Without the Hot Air*, the kilowatt hour is used throughout the book in almost all instances for ease of comparison in grasping the magnitudes involved. It corresponds to the unit of domestic supply, which will cost the householder in the region of 13-14 pence (and rising sharply) and gives a rough idea of the energy needs of a domestic household, since a one bar electric fire will consume in the region of one kilowatt of electricity per hour. On this measurement scale, if every last bit of energy were to be extracted from a kilo of petrol and converted to electricity, it would suffice to keep a one bar electric fire running for 13 hours, compared to 3 hours for a kilo of wood.

The term "bits" of energy skates round an interesting question of dimensional analysis, for the most appropriate units with which to define and quantify energy vary according to the situation. In matters of metabolism or chemical reaction kWh would be a clumsy and opaque unit. Furthermore, while it is assumed that energy can be converted from one form to another, the change is not commutative: that is to say, one can convert electrical energy to heat energy fairly obviously, but how to convert heat - and especially low grade heat - to electricity, directly or indirectly is by no means so obvious. The answer will be seen to lie in using energy to manipulate energy. This is at the heart of the engineering challenge.

Energy equivalence can be expressed in equation form, thus:

$$1 \text{ kilogram metre/sec} = 9.81 \text{ wattseconds} = 2.3 \text{ calories}$$

In words, kinetic energy (strictly, the energy of an accelerating mass) can be converted to electrical energy (by a dynamo or alternator), which can be converted into heat by raising the temperature of water, but it does not appear to be the case that this process can be reversed, nor is it at all clear what kind of stuff, substance or reality is this "energy" that undergoes such changes in form.

## 1.3 Towards a New Energy Management Paradigm

The present work puts forward the case for a particular kind of microgeneration and, as such, its significance can only be fully appreciated by viewing it within the larger framework of electrical generation on other scales. Electricity can be seen to be produced rather neatly on macro-, meso- and micro- scales; that is to say, large, intermediate and small scale. A few words need to be said about their changing relationship before looking at how they can be linked most effectively to resolve the problem of atmospheric carbon pollution.

The most significant changes in electricity generation occurred in the latter half of the nineteenth century, with two revolutions taking place more or less at the same time in the switch from local to central generation and from DC to AC current. In the early years of the age of electricity meso-generation dominated the scene, with each large town having its own independent system and using direct current, until Tesla's pioneering work in electrical transmission persuaded them to convert to alternating current. Microgeneration barely existed, except with some wealthy individuals, who ran their stately houses on home-produced electricity, usually with coal-fired generating systems. The pioneering inventors William Armstrong and Charles Parsons had their own small scale hydro systems.

The change to macrogeneration in Britain was only made possible by the establishment in 1926 by Act of Parliament of a national high voltage grid with standard frequency. This effectively put an end to meso- and micro-generation and was due almost entirely to the efforts of the entrepreneur engineer Charles Merz. He may not have been the first to conceive of a national grid, but it was his lobbying of Parliament for nearly twenty years that made it happen. The present book puts forward microgeneration as such a revolution, but almost in mirror image, for, just as Merz's achievement was to centralize electricity generation, the aim of the proposed E-plus house is to decentralize it, while keeping all the advantages of a national grid. If this can be done – and the looming energy crisis in the UK is an argument that it <u>must</u> be done – micro-, meso- and macrogeneration would work symbiotically. Microgenerating installations would need the Grid to purchase, consolidate and redistribute the many small amounts of electricity which they would produce, and the Grid in turn would call upon the domestic residence as a source of electricity to complement and eventually replace existing coal- and gas-fired power plants. It would be unrealistic to expect micro-generation to be developed on so wide a scale and so rapidly as to entirely do away with the need for centralized generation, and the future mix of micro-, meso- and macro-generation is impossible to predict. Nevertheless, it is important to see them working together in an energy internet for the benefit of society and the planet.

Currently, almost all electricity worldwide is produced by mac-rogeneration, using a national grid fed from a relatively small num-ber of large coal, gas, oil or nuclear fuelled generators. A few places, such as Switzerland and Quebec generate almost all of their needs from hydroelectric generation, and the US generates a significant 10%, but in general hydro is not an option for most countries. Mac-rogeneration from coal, gas and nuclear in Britain is supplemented by some 5,200 giant wind turbines (at January 2014) on-shore and off-shore, which provide just over two per cent of national needs in optimum conditions and almost nothing for long periods. It is widely thought that these will ultimately provide a clean energy so-lution, but the reality is quite the opposite. Another 7,000 wind turbines are planned, but something in the order of 100,000 more wind turbines will be needed, even to make good in theory the loss of conventional power stations due to age or enforced decommis-sioning through European legislation. As a member of the EU, the UK is now legally obliged to generate 32% of its electricity from renewable resources by 2020. Despite the inefficiencies of direct macrogeneration from wind farms, large wind turbines can produce electricity usefully in ways other than by feeding it intermittently from wind farms straight into the National Grid. The most effective in principle would be to combine wind generated electricity with pumped water storage but this does not appear to be part of any present or future plans. Although the energy expended in pumping the water "up hill" is greater than that which is later supplied to the Grid, this drawback is more than offset by being able to bring elec-tricity on stream within a minute when required.

Other forms of renewable macrogeneration, proven in various degrees, are tidal barrage, wave power, geothermal and large scale solar. These receive strong support from advocates of green energy, but their economic feasibility at this stage is unproven. The *La Rance* barrage in France, constructed almost fifty years ago, has proved very successful (not least as a tourist attraction) but partly because the estuary of the River Rance (near St Malo) has some of the highest tides in the world,  up to 13.5 metres. However, despite enormous capital investment its rated 240 megawatt provides only 4% of the electrical needs of Brittany, where it is situated (0.012% of the nation's needs) and, significantly, the French government has no plans to build another, even though the power produced is significantly cheaper per unit that from nuclear installations. The slightly bigger installation at Sihwa Lake in South Korea came on stream in 2011, but in general there are too few appropriate sites worldwide to make this a global solution and, as with dams more generally, silting up is an ever-present problem. Harnessing wave power through oscillating "ducks" or "snakes" is a live area of research, as are tidal and riverine power using submerged turbines, but the slow progress of experimental work (at least as reported in the mainstream press) suggests that their economic feasibility is in doubt. Only one company to date (Pelamis) is at a developmental stage where it can feed electricity to the Grid.

There is growing interest worldwide in the potential of enhanced geothermal generation systems (EGS), with one project well under way in Cornwall, and some forty installations in California already producing 2,400 megawatts of electricity. The process involves breaking up the earth's crust at a depth of about 5 kilometres and injecting water to be heated and recovered to drive steam turbines on the surface. EGS is well proven and economically feasible but is not without snags, as earth tremors in Basel, Switzerland, and Anderson Springs in California have demonstrated. In these and other places EGS projects have had to be abandoned. Use of solar energy on the large scale is developing at a speed that makes it hard to keep up with and will certainly play an increasing role in the future. In desert conditions, from the Arabian peninsula to central Australia and California installations of different designs are springing up. All depend at base on harvesting the sun through hectares of mirrors focusing it to create very high temperatures and using the concentrated energy to generate electricity through turbines and generators.

Mesogeneration goes by several other names – e.g., embedded, on-site, or decentralized generation - but is usually referred to as distributed generation, or DG for short, a term made popular by Jeremy Rifkin, who called it "the third industrial revolution".[5] DG refers to a system half way between the top-down, centralized macrogeneration and bottom-up microgeneration, which is the main concern of this book. The output of a DG system is typically between 100 and 2,000 kilowatts, at the bottom end servicing, say, a small factory and at the top end a small village. The term "embedded generation" indicates that it exists within and as a part of the national grid, and this creates a variety of problems, on which many governments and power companies have been working, such as the stepping up or down of voltages and the smoothing of peaks of supply and demand. There are many examples of DG in operation in several European countries, with the Fraunhofer and Dardesheim projects in Germany figuring prominently, and already serving communities of small town size and universities like Darmstadt concentrating research into this particular field. Among other examples are the Charles Dickens estate project in Portsmouth, the Drake Landing Solar Community in Alberta, Canada, the Anneberg solar project in Sweden and the Wiggenhausen-Süd in Germany. The latter three are of interest in their different strategies for thermal storage, respectively by using deep boreholes in the earth, crystalline rock and extremely large concrete water tanks (12,000 cu. metres).

As against the K-gen/E-plus concept, which sees microgeneration as a primary source for the Grid, DG initiatives tend usually to seek independence from the Grid as an ideal. Thus if the DG movement were to spread, we could see a return to the kind of fragmented and disconnected distribution system that existed before the National Grid was set up by act of Parliament. Another and major difference is that reduction of atmospheric pollution is not always a prime consideration in DG complexes, for while they make use of solar and wind energy, which are clean renewable sources, they often also seek to serve a public need by disposing of waste through incineration, thus releasing carbon dioxide into the atmosphere. Furthermore, most aim for efficient use of natural gas, which is neither clean nor renewable. Their main focus is not on reducing global warming, they tend to think and plan locally and not within the context of national energy requirements. Also, and importantly, while they sometimes feed electricity back to the grid, this extra supply is not only sporadic but is mostly available when the grid has spare capacity. Awareness that DG has a new role to play is growing rapidly, and the Institution of Engineering and Technology says quite categorically that "a range of different generating technologies, some qualifying as DG, will develop in the future driven by fuel costs [and] environmental pressures" and these are "challenging the existing architecture of the networks."[6], which may be a euphemism for putting impossible strain on the Grid.

Decentralizing electricity supply through micro- and meso-generation has two great advantages, the first being that macrogeneration from fossil fuels is immensely wasteful, as the diagram below illustrates. Between the loss of heat energy in cooling towers and the loss of electrical energy in transmission, hardly more than a third of the energy in the molecular bonds of coal is available for use. That really is a scandalous figure.

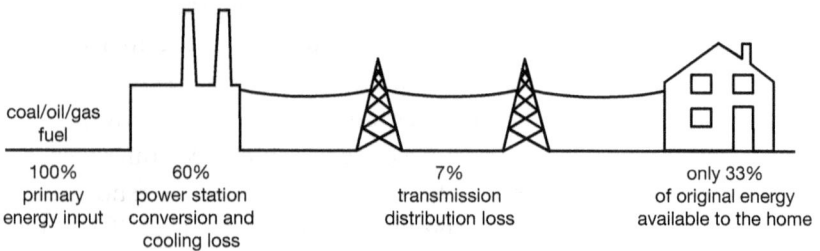

| 100% primary energy input | 60% power station conversion and cooling loss | 7% transmission distribution loss | only 33% of original energy available to the home |

Fig. 1   Energy loss in macrogeneration

Grid design is a highly technical problem, but of the alternatives - based on fossil fuel, wind power and microgeneration - it appears that the existing Grid can be more easily, cheaply and rapidly adapted to microgeneration. Government policy, however, is going in exactly the opposite direction, towards a reconstruction of the National Grid with a view to maximizing input from wind farms, regardless of the massive case against them that is building up. Amid the tales of inertia and bad decision-making which surround the whole issue of sustainable energy in the UK, it is encouraging to note that the Institute of Engineering is a very active participant in Innovative Smart Grid Technology (ISGT), an international body, and hosted the 2011 European conference and exhibition in Manchester. In the dozens of papers and poster presentations on all aspects of this, such as voltage control and load shedding, there was recurrent awareness of distributed generation, but despite one paper entitled "The Smart Home", microgeneration seems to have received little attention. Given the Desertec Initiative, funded by the World Bank, which aims to bring solar-generated electricity from north Africa through to Europe in huge quantities within fifteen years through a continent-wide grid, there is no reason to think that much attention will be paid to microgeneration.

The second disadvantage of macrogeneration, and one that is rarely considered, is the vulnerability of high voltage and long distance transmission lines to total breakdown from solar eruptions. This has happened and inevitably will happen at some point in the future. Solar storms do not normally attract general attention, but actually constitute a massive threat to grid-distributed electricity. The head of NASA's Heliophysics Division, Richard Fisher, has predicted:

> In the next few years we expect to see much higher levels of solar activity. At the same time, our technological society has developed an unprecedented sensitivity to solar storms .... We know it is coming but we don't know how bad it is going to be. It will disrupt ... the banking system, our computers, everything that is electronic. It will cause major problems for the world.[7]

He should have also said that it will disrupt everything electrical as well as electronic. In 1859, at the birth of our electrical civilization, a solar storm shorted out telegraph wires in Europe and America, starting many fires. In 1921 a similar "plasma strike", which created violent surges in the grid, knocked out the New York transit system. More recently, in 1989 the Quebec power system was crippled. Such geomagnetic storms are accompanied by dramatic increase in the intensity of the aurora borealis, making it visible as far south as Rome and even Cuba. A 2009 *New Scientist* article explained the danger thus:

> *The incursion of plasma into our atmosphere causes rapid changes in the configuration of Earth's magnetic field which, in turn, induce currents in the long wires of the power grids [which] were not built to handle this sort of direct current electricity. The greatest danger is at the step-up and step-down transformers used to convert power from its transport voltage to domestically useful voltage. The increased DC current creates strong magnetic fields that saturate a transformer's magnetic core. The result is runaway current in the transformer's copper wiring, which rapidly heats up and melts. This is exactly what happened in the Canadian province of Quebec in March 1989, and six million people spent 9 hours without electricity. But things could get much worse than that.*[8]

There are reasons for thinking that the next great solar storm will be upon us within the next five years, and the damage could be immense. Fisher estimates that without radical and expensive alterations to the world's grid systems, it could take four to ten years to recover and could cost as much as $2 trillion dollars in the first year alone. Against this background it can be seen that mesogeneration and, even moreso, microgeneration will bring an enormous advantage, since they can function as independent units even if the grid itself is put out of action, though they would normally be grid-connected. They are relatively immune from damage because their low voltage, low capacity wiring does not require the complex system of transformers which is the most vulnerable part of a centralized grid. They could be, in fact, the back-up system that Fisher calls for.

## References

1. Marion King Hubbert, from an interview with Robert Dean Clark, "King Hubbert: Science's Don Quixote," *Geophysics* Magazine, Feb. 1983.

2. George Monbiot, *Heat: How to Stop the Planet Burning*. London: Allen Lane, 2006. p. xi.

3. Arthur Eddington, *The Nature of the Physical World* [the 1927 Gifford Lectures]. CUP, 1933. pp. 74-75.

4. The systematic disinformation is documented in N. Oreskes and E. M. Conway, *Merchants of Doubt: How a Handful of Scientists Obscured the Truth on Issues from Tobacco Smoke to Global Warming*. Bloomsbury, 2011. A review in *Science*, urged that "all those engaged in the business of conveying scientific information to the general public should read it." One wonders, however, what national daily has enough public spirit to take the risk of revealing these commercial black ops which affect all citizens.

5. Rifkin's point is, in brief, that the first industrial revolution began with the availability of energy in the form of coal and the invention of the steam engine, the second with the discovery of oil and the invention of the internal combustion engine, and the third is now being ushered in with the use of hydrogen as a non-polluting fuel source to replace oil, gas and coal, with generation through meso-scale systems integrated into national or regional grids. Rifkin is a visionary thinker, and his 2002 book *The Hydrogen Economy* is subtitled *The Creation of the Worldwide Energy Web and the Redistribution of Power on Earth*. However, there are serious problems with hydrogen, analysed at length by Joseph Romm in *The Hype About Hydrogen: Fact and Fiction in the Race to Save the Climate* (2005). Romm is eminently qualified to speak, having been an adviser on energy policy to the Clinton administration. The usefulness of hydrogen as a clean fuel in microgeneration will be taken up later.

6. Institution of Engineering and Technology Factfile. See "Distributed Generation" on the Internet.

7. http://science.nasa.gov/science-news/science-at-nasa/2010/04jun_swef

8. Michael Brooks, "Gone in 90 Seconds", *New Scientist*, 21 March 2009. p.32.

# Chapter 2

## Peak Oil and a Choice of Disasters

*There is not a nation on earth that is preparing intelligently for the end of oil.*

James Howard Kunstler, author of *The Long Emergency*

*Avoiding dangerous climate change is impossible – dangerous climate change is already here. The question is, can we avoid* **catastrophic** *climate change.*

David King, UK Chief Government Scientist, 2007

### 2.1  Peak Oil: The End of an Era

The global energy crisis is a complex matter, with different kinds of disasters looming at different levels. The first and most obvious fact is that the world has finite supplies of oil, which must at some point be used up. Over the past ten years or so this situation has become widely acknowledged in the term "Peak Oil", expressing the fact that global supplies of conventional oil are now being used up faster than new oil fields are discovered. There are obvious uncertainties, but superimposed graphs of falling supply and rising demand show two critical facts: the curves crossed about 2008, meaning that we then started to use up conventional oil faster than new sources were being found. Since then new sources have been found in fracked shale oil and gas, and new oil recovery techniques, mainly horizontal drilling and so-called "water flooding", have greatly improved the most pessimistic forecasts. They do not, however, alter the overall picture, but only move the crossover point forward into the future, perhaps to 2018, or at best 2030. Forecasts must be uncertain, for there are many variables and unknowns, but it is reasonable to foresee that 90% of the world's oil will be used up by 2070, give or take ten years. One obvious variable is conventional oil from sources as yet undiscovered, another the amount from unconventional sources such as tar sands, now being exploited in Canada and Venezuela. The problem with the latter is that recovering the oil calls for huge expenditures in energy and water and creates equally huge pollution. The EOEI yardstick (energy obtained over energy invested) decreases continually. Oil from the early "gushers" in the US, Persia and Saudi Arabia typically produced a hundred times more energy than was expended in the drilling procedure, whereas oil from tar sands and fracked shale formations typically gives back only three or four times the amount of energy required to extract and refine it. The cost in environmental devastation is, unfortunately, unquantifiable and never included in the oil majors' annual reports.

Forecasts for future oil supply must balance the unconventional sources now being developed against the diminishing supply from old sources. About a quarter of the world's energy supply comes from so-called "giants", that is, oil fields which can steadily produce in the region of 500,000 barrels per day. There are about 25 such fields but most of them are seventy years old and many are depleting much faster than predicted, not least the North Sea field.[1] Given these uncertainties, we may expect the global oil supply to halve by about 2030-2040 and the price of oil - and everything dependent on it - to rise accordingly, sending up the price of almost everything in its turn. Anything which must be transported will have to rise in price to offset increased fuel prices.

When we unpack the term "everything dependent on oil", the picture that emerges presents a daunting challenge. We in the West have lived for the last century in a completely anomalous era because of the abundance of cheap fossil fuel which will no longer be available well within the lifetime of our children. The word "West" has become more or less co-extensive with the ordinary citizen's ability to flip a switch to obtain light, turn the ignition key in a car to have private transport, turn on a tap to have clean and abundant water, drive to a shopping centre where one can obtain food and goods from all over the world or book a ticket online to fly a thousand miles for a weekend break. We take this cornucopia for granted, but a little more than a century ago it would have been fantasy. On television one can watch container ships, powered by diesel engines the size of house, shifting vast quantities of goods across the planet, but when my father was a child, it was nothing unusual for him to watch sailing ships unloading cargoes of timber and cotton in the seaport where he was born. We have been utterly profligate in the way we have used oil, and still are. A current project in Kuwait to create a Venice style residential area with manmade lagoons has used bulldozers and huge dumper trucks to move sand and silt in quantities enough to fill Wembley stadium 67 times over. To fill the fuel tank of a 400 tonne giant dumper truck for a day's work costs about £5,000. That thoughtless waste of oil must eventually come to an end, and when it does, we will no longer be able to take for granted many of the just noted aspects of modern civilization.

The precise time that it will come to an end is almost irrelevant when considered against what will happen thereafter. The changes in global civilization that must ensue are too great to be imagined in any systematic way, and though it may be hardly possible for even the wisest of think-tanks to get a grip on the problem, one might at least get a sense of its enormity by seeing the picture as a reversed image, as in a photographic negative. Undeveloped countries, what used to be called the third world, have long aspired to Western living standards, having seen in the cinema another world in which everyone has a car, a spacious house, big fridge and abundant food, but if only a half of the population of Africa, India and China were to possess a car, the whole of the planet's oil supply would be swallowed up in twenty years. In this imagined, and fantasy, scenario we would then be left with a world without energy to run anything – unless, that is, we can organise society in such a way as to make renewable energy our main source, and stop viewing it as a footnote or substitute. There is simply no hope of a world population realising what is often referred to as "the American dream", and the dream itself is slowly turning into a kind of nightmare, with fifty million citizens of the once richest country in the world now only surviving thanks to food stamps. Only when this changing situation is fully realised will the significance of the K-gen and E-plus initiatives be appreciated.

It is when we come to look at the products of oil, however, that we realize how great an impact the discovery of oil has made on our lives. One such assessment lists 160 items of which petroleum is the feedstock which will either be unobtainable or much more expensive when the oil runs out.[2] Among the items listed are:

*Antiseptics, aspirin, car tyres, dentures, detergents, fertilizer, floor wax, footballs, garbage bags, glue, golfballs, guitar strings, heart valves, luggage, lipstick, nail polish, nylon rope, paint, paint brushes, parcel tape, petroleum jelly, roofing felt, shampoo, tarmac ...* and on and on

No doubt the science of chemistry will find other feedstocks from coal and fibreglass, and the potential that lies in pyrolizing biomass is the subject of Appendix I, but with a rapidly increasing global population, it will hardly be possible to satisfy everyone's needs. Reading this very partial list brings home how petroleum has transformed our lives and created new lifestyles in many ways. It has brought a new civilization into being, whether or not we may think that its negative aspects outweigh the positives. This civilization built on unlimited cheap oil is now passing away, and we do not know what will, or could, replace it. Matthew Boulton typically clinched the sale of Watt's steam engine with the line, "I am offering you what all mankind desires - power," a sales pitch reminiscent of the one that the devil made to Goethe's Faust, and we can see now that it was indeed a Faustian pact. For just as a dreadful day of reckoning came when Faust's natural life of unlimited power came to an end and the devil claimed his soul, the inevitable exhaustion of fossil fuel will bring with it an accelerating collapse of the civilization created with its power. As global oil supply diminishes, slowly but certainly, a major restructuring must take place, a downsizing and shrinking on a scale that will change civilization. As earlier noted, cheap energy has given us all the equivalent of dozens of slaves, and enabled us to construct a complex society, with whole sectors, such as education and health care paid for by this hidden source of labour and a largely unnoticed economic infrastructure laid down.

How will we, for instance, maintain our road transport system in the future, when the national wealth that flowed from cheap oil is cut back and many people, probably the majority, can no longer afford to run a car? It is difficult for most people even to imagine that as the supply of the world's oil comes to an end, whether it be in fifty or a hundred and fifty years, the raw material for road-making will run out. Tarmac is a complex substance with a complex history, but is composed essentially of fine gravel and bituminous tar, which is sometimes found in natural deposits (as in Trinidad) but is mostly a by-product of oil refineries, otherwise wasted. When the tar is no longer available, how will future generations build or maintain highways? It is possible that chemical engineering will enable an improved coal-based tar to be used as a partial substitute, but even if that turns out to be the case, the pollution problems associated with coal will remain and get worse. In the past hundred years, an astonishing network of major and minor roads and urban streets has been constructed across the globe from oil-based tar. These millions of miles of transport infrastructure must eventually shrink, as if winding back a film of their creation, and what will remain as a permanent road system is hard to visualize. An honest answer must be that we shall simply have to let the highway system disintegrate selectively, as is already happening by considered decision in some American states.[3] Such a situation, now exceptional and ad hoc, will almost certainly be a matter of policy in twenty years and is but one consequence of the global energy crisis. Some modern highway surfaces have been deliberately designed to be recycled through being periodically "planed", melted and relaid, with ingenious and energy-hungry road-making machines. However, while statistics are not available, this is unlikely to account for more than 20% of the total and more probably 5%. There is a great temptation to say that "science" will take care of the problem, but without the raw material that has gone to creating the road network on which we depend, that looks like an unjustifiable act of faith from the present standpoint.

For some 150,000 years growth of the earth's population was very slow, largely because lack of proper nutrition and health care made life expectancy short - typically 45 years. The astonishing increase in the past 150 years has only been made possible through a revolution in medical science and explosive growth in world food production. The latter has been brought about by the use of artificial fertilizers and pesticides made from natural gas and petroleum, and from pumped irrigation, which now accounts for almost 20% of US farm energy use. As fossil fuel energy becomes ever more scarce and expensive, there must come an exponential decrease in the world's food supply, and this will occur as population growth is still increasing exponentially. The situation will almost certainly be made worse by the use of agricultural land to grow crops for biofuel rather than food in order to make up for a shortfall in petrol for transport. Already in the US 14% of corn (maize) is grown for ethanol production, which is hugely wasteful in energy terms, but government subsidized.

The connection between energy and population adds a critical new dimension to the Malthusian thesis that unregulated population must eventually outstrip resources, but only a few and recent analyses emphasize the point that it is only abundance of fossil fuel that has enabled the population to increase in the uncontrollable way that it has.[4] Agriculture is the single largest consumer of petroleum products, even greater than transport and electricity generation. It accounts for 17% of gross demand and one can expect to see world food production fall by a comparable, or even greater, amount as world population continues to rise.  It is probably true to say that (China apart) nothing has been done about overpopulation at national or international level. It is indeed very difficult to see what can be done, but simple logic would tell us that within our children's lifetime there will be famine on a scale that history has never seen and is barely imaginable. Without doubt this will set off mass migrations, as famine has done throughout human history, with all the dire consequences for large scale violence that must follow. Those who would scoff at such an apocalyptic scenario need only to look at the graphs. The onus of proof is on those who believe, or hope, that the world can return to a past now disappearing. Logic is on the side of those who argue that we must seek and adapt to a new normality.

## 2.2  Global Warming and Climate Change

Climate change is happening through global warming, which in turn is happening through atmospheric pollution. The main natural causes of this pollution are forest fires, volcanoes and methane release, whereas man-made pollution is created almost entirely by carbon dioxide emissions from industry, transport and domestic energy use. The UK domestic energy requirement is about 25% - 30% of all the nation's energy needs, and hence of the nation's atmospheric pollution. Thus if every house in the UK could meet its own energy requirements from renewable and non-polluting sources, atmospheric pollution from burning fossil fuels would be reduced by a comparable figure, and if all houses could on average steadily generate twenty per cent more electricity than they needed, which is certainly within the bounds of possibility, there would be a surplus of electricity for industry. But the benefits go further than this, for if this figure were proved feasible at the level of microgeneration, and if it could be applied to mesogeneration, there would in theory be no need at all, or hardly any need, for macrogeneration through coal-fired, gas-fired or nuclear stations. This is, of course, an ideal scenario, but the important thing is that it is not impossible in principle, and it would certainly appear more realistic when the number of domestic units earning money through surplus electricity reach a critical mass. Finding a non-polluting source of energy for transport is a separate and less tractable problem, and not within the scope of this book, but the potential for a methanol or ammonia based fuel strategy will be touched upon in Appendix II.

However successful the theory of microgeneration and its implementation may turn out to be, atmospheric pollution will continue unabated for a long time to come simply because electricity demand will continue to be filled by burning fossil fuels. To give some idea of the figures, the US generates 49% of its electricity from coal, and China 70%, and greater use of alternative fuels, more particularly natural gas and uranium, will not substantially affect these figures. In 2008 China consumed 2.62 billion tonnes of coal, and is constructing almost one new coal-fired plant a week. In 2007 the world had about 1,300 gigawatts of coal-fired electricity generation, and the International Energy Agency predicts that this will grow to over 2,000 gW by 2030.[5] These brief statistics may serve to put the global problem in perspective. As in all such forecasts made prior to the global economic depression now developing, the figures must be taken as speculative. They will, however, serve as initial evidence that we are entering into a period of greater pollution and in all probability runaway climate change.

The devastating effect on our species and civilization of continuing down this path stretches imagination to the limit. A plot of atmospheric carbon dioxide concentrations over the past 200 years fits with uncanny exactness on a curve showing the use of fossil fuel. The first practical steam engine, which initiated the industrial revolution, came into use in 1798, and the graph clearly shows global warming from about a century later, with the average temperature rising since 1900 by perhaps an eightieth of a degree Celsius per year. The 19th century is the flat part of a curve that is now starting to turn towards the vertical. [6] The process has been both irregular and concealed by the cooling effect of some short-lived particulate emissions and various climatic cycles, so that the steepness of it has taken most forecasters by surprise, causing them not only to constantly revise predictions but dramatically so. The figures are even more confusing, since the last ten years have seen no atmospheric warming, though this may well be due to the oceans absorbing the surplus heat, for reasons that are as yet unclear. As long ago as 2000, the Intergovernmental Panel on Climate Change (IPCC) Special Report on Emissions Scenarios drew attention to the fact that before the Industrial Revolution there were 280 carbon dioxide particles per million in our environment, which has risen today to 380 and is projected to be a minimum of 540 by the year 2100. We do not know what the effect even of this minimum will be, but already we are seeing in some places die-back of coral reefs through oceanic warming. That surely must warrant a worst case assumption.

In addition to the evidence from oceanology and satellite observation, there is a wealth of data on global warming and climate change from ground based records and many expert reports. Of the easily available literature, the Stern Report of 2006, commissioned by the UK Treasury is a good place to start in defining the problem and getting a sense of the obstacles in the way of a solution. In the first place, the report was entitled *The Economics of Climate Change*, thus diverting emphasis away from the ecological damage arising from global warming and onto the short term economic effects of any proposed solutions. Even as early as 2003 among effects catalogued by the World Meteorological Organisation were the hottest June in Switzerland for at least 250 years, and 562 tornadoes in one month in America, against the previous high of 399 The WMO notes that "new record extremes occur every year somewhere in the globe, but in recent years the number of such extremes has been increasing." We are yet to see what happens when these isolated cases become part of a new pattern of global weather, but one small example from Canada may serve as a minor illustration. The forests in British Columbia have been increasingly ravaged by the pine beetle, which has destroyed about 40 per cent of the pine and is now attacking spruce trees, previously unknown. What has been found is that in the past the beetle had been kept under control because most of its larvae had been killed off by winter frost, but with average winter temperatures rising by only a degree or two, the survival rate has risen to pest proportions.

A great deal of new and alarming data has appeared since the Stern Review, the most significant being the discovery that the process of global warming is not only a positive feedback loop but a complex system in which several reinforcing loops accelerate change hyperbolically. This was made clear in a report entitled *Planet Earth, We Have a Problem: Feedback Dynamics and the Acceleration of Climate Change* given to the Parliamentary All-party Climate Change Group in June 2007.[7] The thrust of the paper is that global warming is not only accelerating, but the rate of acceleration is increasing, and forecasters are thus chasing a moving target. The second order feedback of "radiative forcing" is generated from emissions of carbon dioxide, decreasing reflection as the polar ice caps shrink, making almost all previous forecasts out of date and misleadingly optimistic. The large and uncertain number of factors in a positive or negative feedback relation is a challenge to the most powerful computers or sophisticated logicians, and this gives some specious cover for so-called "climate change deniers". There was, for example, a levelling off of atmospheric methane for thirty years towards the end of the last century, probably due to changes in agricultural practice, but at the same time good reason to think that future thawing of the sub-Arctic permafrost could release devastating amounts of methane into the atmosphere.

What little uncertainty there may be about the urgency of the crisis has been dispelled by additional hard data since the Westminster Briefing paper was presented. The 2009 annual meeting of the American Association for the Advancement of Science reported that satellite measurements now call for radical re-evaluation of global warming and rising sea levels, which it says is "now outside the entire envelope of possibilities considered in 2007." More recently a joint report by the UK and US Met Offices argues "unequivocally that the world is warming and has been for more than three decades."[8] Against this background a programme of carbon credits exchange is not only an irrelevance, it is an insult to the intelligence of the average person.

Unless humanity can organise itself to take massive countermeasures, the simple outcome of a very complex process is that the earth's climate will flip into a new state, with dramatic changes in the world's flora and fauna. There are indeed well qualified commentators who foresee a sixth "extinction event" ahead as the end state, like the one that wiped out the dinosaurs. The "good" news is that it is unlikely just to get warmer until the planet burns up and life disappears, but will more probably stabilize at five or ten degrees hotter, when smoke from forest fires, increased cloud cover and high altitude water vapour, and perhaps other aerosols, have a damping effect. We are now in a race against time to slow down this process and eventually halt it.

## 2.3 The Mathematics of Disaster

In an area of great uncertainty and potentially great disaster, two immediate needs become apparent – for political action and for the firm data that will justify radical action. While there was never a more urgent need in human history for wise and decisive political leadership, there is, sadly, at every of level of government a deficiency that it seems we shall have to accept. The problem is planetary, but jurisdictions are territorial and based on the nation state. So no one is really in charge. Furthermore, the form of government that we now take for granted, namely, party political democracy, is heavily biased against long term solutions. Against the ideal of democratic government, the reality is that the world, in the West at least, is governed by individuals who, at best, put national interests before global needs and, at worst, make their own political benefit and that of their party a priority. No one with political power speaks for the planet or for future generations. This chapter is headed by a quotation from Sir David King, the official scientific adviser to three governments, and a tireless advocate for a rational policy to avoid the worst effects of climate change, but there is no evidence that any of the three main political parties have taken any notice. A similar situation exists in the United States, where James Hansen, who for many years headed NASA's satellite global temperature surveillance, has resigned to devote all his time to raising public awareness of the threat of climate change.

Hansen, often vilified as an extremist, deserves particular attention, for his expertise also lies in climate change on other planets, notably Venus and Mars, which once had far more temperate climates than the present. His message is, in a nutshell, that given an excess of greenhouse gas, such as carbon dioxide, a once relatively benign climate with water vapour and oxygen as a source of life can become totally barren. His argument is that beyond a certain point of atmospheric degradation there is nothing to stop planet earth following the same path and ending up in a similar situation to Mars. In his words, "I've come to the conclusion that if we burn all reserves of oil, gas and coal, there is a substantial chance we will initiate the runaway greenhouse. If we also burn the tar sands and tar shale, I believe the Venus syndrome is a dead certainty.[9] It is by no means clear how many centuries such a development would take, but this is the trajectory if we do not soon take urgent and systematic action.

The data that is needed to persuade politicians and people in general that radical action is necessary has been presented briefly and starkly by the environmentalist Bill McKibben.[10] He reduces it to three critical and easily understood numbers, namely two degrees

Celsius, 565 gigatonnes and 2,795 gigatonnes. The connection between them and its significance is as follows. The 2009 Copenhagen Conference on climate change, which Sir Nicholas Stern said should have been "the most important gathering since the Second World War", failed to address any of the issues in a serious and decisive way. Despite high words, it ran away from the problem of atmospheric pollution and was, as one commentator has put it, a latter day Munich. One of the few things agreed was that the world could not sustain more than a rise in global temperature of more than two degrees above the historical norm. Even this agreement could be interpreted as kicking the problem into the long grass, for there were some expert advisers who pointed to the observable effects of a rise to the current level by 0.8 degrees, notably an increase in sea water acidity of 30%, the loss of a third of the Arctic summer ice and a 5% increase in the humidity of the above-ocean air. A few moments thought is sufficient to realise that the latter amounts to a huge amount of additional water in the atmosphere that has eventually to come down, and has already resulted in more frequent and more powerful hurricanes and unprecedented and unseasonal rainstorms followed by devastating floods. In the face of this evidence, there were those who warned that an increase of only one degree would be a huge gamble, but for the sake of putting some sort of international agreement in the final report, a compromise figure of two degrees was accepted.

What McKibben added to this were the second and third figures which had been reached from the best available data and the most powerful computers. 565 gigatonnes was the amount of carbon dioxide that the world could, as it were, allow itself to release into the atmosphere before the critical increase of two degrees was reached. There was every reason to think that some sort of climatic breakdown would have happened before that, but this figure represented a limit beyond which optimism must become self-delusion. The third figure of 2,795 gigatonnes summarises the global energy trap, for it represents a best estimate of the amount of carbon dioxide that would be released into the atmosphere if all the world's remaining fossil fuel reserves were to be used up right to last drop or shovelful. It is, of course, artificially precise, since useable reserves are best guesses, but likely to be an under-estimate, as new discoveries in shale gas and oil are changing the picture.

It is important to realize that McKibben's projections are by no means the most pessimistic. Kevin Anderson, Professor of Energy and Climate Change at Manchester University, is prominent among those who interpret the same data to predict that the world is on a path towards potential 4C rises as early as 2060 and 6C by the end of the century with consequences that will be, in his words, "terrifying". "We will not make all humans beings extinct, as a few people with the right sort of resources may put themselves in the right parts of the world and survive. But I think it's extremely unlikely that we wouldn't have mass death at 4C. If you have got a population of nine billion by 2050 and you hit 4C, 5C or 6C, you might have half a billion people surviving."[11]

It is difficult to comprehend the scale of the threat posed by global warming to the planet and the human species. Not only are the data complex and often ambiguous and best and worst case scenarios widely divergent, but each part of the planet will suffer its own particular consequences. As the oceans and atmosphere continue to warm, the effect may range from more frequent and more violent winds to drought and destructively heavy rainfall or, as weather patterns shift, to failure of the monsoons. Every country will suffer from a different mix of these effects. A recent book *Four Degrees of Global Warming: Australia in a Hot World* is a particularly valuable reference because Australia is not only one of the world's most civilized nations but a continent. It is worth noting that whereas 4° was until quite recently regarded by most climate change specialists as a worst case scenario, it is now considered a median figure, with some anticipating a rise of 6° a possibility. The Australian study argues that there will be "dramatic climate change" if carbon emissions are not cut by 40% before 2020. Short of putting the country on a war footing (as Jeremy Leggett calls for in the UK) anything like that target is completely unrealistic. However, when the effects of 4° warming are felt, the need for national mobilisation will be clearly seen, for Australia is among the most ecologically vulnerable to climate change. In the first place, a rise by 4° predicted by 2100 would effectively destroy the Great Barrier Reef, as shallow ocean temperature rises and the coral dies off. The Murray-Darling river system, which is a key element in Australian agriculture, and already near its limit of stress from over-demand, would dry up so catastrophically as to wipe out 90% of the food production now dependent on it. Several major cities would endure summer temperatures at a level normally considered to be unfit for human habitation – in effect, a two month heat wave with daily temperatures in the 35-45° range. (The figures have been simplified but not exaggerated.). As the editor, Peter Christoff, says, Australia's future is a "disturbing and bleak vision of a continent under assault."[12] It would not be untrue to say that global warming puts the whole planet under assault.

Long before fossil fuels are exhausted the global atmosphere will have been wrecked and an irreversible trend set in place. The former leader of the Green Party, Caroline Lucas MP, in a letter to *The Telegraph* (14/08/13) argues that the tipping point will come unless we leave 80% of all fossil fuels in the ground. That is the figure reached from analysis by Carbon Tracker. As earlier intimated, and as Wadhams, Hansen and others have emphasized, if the worst predictions of methane release in the frozen tundra are realised, no human action, however drastic, will be enough to stop runaway global warming. It is, in fact, arguable that we have already reached the tipping point, for a report of the World Meteorogical Organization points out that July 2012 was the 329th consecutive month of above-average global temperatures and the hottest month ever for North America.

Even a complete cessation of burning fossil fuels would not have the slightest effect. From what has just been said about political inertia, it can be deduced that non-governmental action must be encouraged as a matter of urgency, and this is where the significance of the K-gen system and the E-plus house comes into focus. At this point they are in conceptual form, but they give reason for thinking that both of them, and the K-gen system in particular, can deliver all domestic and industrial energy needs from clean, carbon-free sources. Additionally, some initial evidence will be provided later to suggest that carbon dioxide levels can actually be reduced by a coordinated strategy to sequester atmospheric carbon by pyrolysis. The overarching point is that while the latter must depend on political initiatives which, as just noted, are effectively paralysed so far, if the K-gen system can be developed to a point where it generates electricity cost-effectively on the microscale, no other inducement will be needed. Ordinary citizens will rush to take advantage of such a profit-making device. Taking the suggestions outlined in this book to proof-of-concept stage thus becomes a clear and urgent task. There will, however, be resistance from those who, for psychological reasons, believe that global warming is some kind of green conspiracy and from those institutions which would lose their profits if sales of fossil fuel were to diminish or be taken away altogether.

### 2.4   Global Warming and Big Business

Despite all the evidence, many otherwise intelligent and objective people remain to be convinced that global warming is actually taking place and maintain that it is more rational to interpret it as simply part of a short- or medium-term cycle. Hence, they conclude, there is no need for action. Interpreting the data may indeed be complex, and summaries may alarm by oversimplifying and over-dramatizing. Nevertheless, history shows that destructive climate change does happen even without human intervention. In the Neolithic Subpluvial phase between about 7,000 and 3,500 BC, sometimes called "the green Sahara" period, the north of Africa was rainy and fertile, and was the homeland of the Semitic people, including the Jews, who were forced northward to Palestine and the Middle East in order to survive, as rain patterns changed and eventually ceased. Hard as it may be to imagine, archaeology has turned up evidence in the Sahara that there was once flourishing agriculture where now there are only endless sand dunes. This at least should give the sceptics pause.

Many of those who wish to avert their gaze from the possibility of climate change have justified their scepticism by reference to the distortion of data revealed in the hacked emails from the University of East Anglia Climatic Research Unit, but it should be remembered that this was not the work of some truth crusader randomly surfing the Internet, but part of a well orchestrated operation, whose findings were very purposely given to the media, thus leading one to ask why it was done and on whose behalf. The obvious path to an answer is to ask *Cui bono?* - who would stand to gain by lulling the community into a false sense of security by making global warming look like some kind of conspiracy theory? - and the trail of investigation leads to the gigantic transnational companies which sell fossil fuel. There is no question that a long-established and well-funded institution exists to manipulate public opinion with methods that may accurately be described as black ops. [13]

There are perhaps two or three centuries reserves of coal available (much in America) when the oil runs out, and the discovery of shale oil and gas in the US and elsewhere, plus giant new gas fields in southern Iran and further east promises equally enormous profits to those companies and nations that control the production of coal, oil, gas and the pipelines and tanker fleets already being built to sell LPG (liquefied petroleum gas) to Europe, China and India. At stake here are future trillions of dollars, and it would be naïve to think that those corporations and individuals who risk losing them would stand passively by and let themselves be dispossessed by a coordinated renewable energy programme.

The role of governments and big commercial interests in all this is ambiguous. On the one hand political decisions are needed at national level in order to coordinate a global policy to reduce global warming, but national interests are almost always going to trump global needs, and commercial interests will trump both. While transnational companies which make their profit from selling polluting fuels will not want to pass up an extra source of profit from renewables, they will never pursue this at the cost of making their existing and hugely expensive infrastructure redundant. The oil majors, particularly Shell, Texaco and BP have been devoting substantial amounts of capital to developing solar, wind and hydrogen energy generation, though there appears to be no comparable initiative from the large coal and natural gas companies, but for a long time to come all must take their profit in the usual way from selling as much of their existing product as they can. Business is business, and the future of the planet is someone else's business.

The blocking role of big power companies in a clean air policy is much more than foot-dragging. The Clean Air and Security Act (ACESA) passed in July 2009 by the American Congress as a response to climate change is an outrageous example of the hijacking of the national and global agenda by special interests. Despite President Obama's declaration that it is "a vote of historical proportions … that will open the door to a new, clean energy economy," the reality under the reassuring rhetoric may be gauged from the single fact that the act waives all controls on "Big Coal" until 2025. As the response to ACESA from all ecologically committed bodies makes clear, government policy has not so much been subverted by fossil fuel interests as set by them, Friends of the Earth says about ACESA: "Corporate polluters including Shell and Duke Energy helped write this bill [which] eliminates pre-existing EPA authority to address global warming and is actually a step backwards." Social Ecology calls it "politics-as-usual while the planet burns." Public Citizen's response was, "This bill is a new legal right to pollute … that will enrich already powerful oil, coal and nuclear power companies."[14] This is the reality of the measures that the United States will take to combat climate change.

If one fact were needed to show how top-down change is blocked by short-term business and national interests working in harmony, it is that since the Kyoto Protocol was first agreed in 1997 there has been no noticeable diminution of air pollution. Most countries and most big companies seem to be set on evading what trivial adjustments were agreed upon and it is difficult to see what legal teeth could be put into enforcing compliance on a global scale. If further evidence were needed that the carbon exchange "market" is in reality a rigged casino, the JP Morgan bank has put in charge of its trading operations Blythe Masters, the financial engineer who invented credit default swaps. More than anything else it is the ingenious derivatives which she introduced to a predatory banking system that have brought the global financial market to its knees and have been called by Warren Buffett "weapons of financial mass destruction".

The same destructive intelligence is still at work in the field of climate change.

From the brief and condensed information in this chapter it can be seen that the global energy trap is complex, but clear to anyone who is prepared to think about it. The world is running out of the fossil fuels which have enabled civilization to make such dramatic advances over the past couple of centuries. Ample and cheap energy has been a vital ingredient in creating modern culture and is essential in maintaining it. Without energy such as we have come to expect we may expect cultural regression. As the world's supply of oil diminishes and the price will rise, economic activity will slow and finally come almost to a halt, unless it can be replaced by a clean, natural supply. At the same time, if we persist in using up fossil fuels much longer, rather than finding a clean and renewable substitute, we shall bring on global warming, with consequences that will be apocalyptic in the most literal sense. We shall take the green and pleasant planet that we have inherited and turn it into a wasteland.

### References

1.  A 2011 report by the International Energy Agency reports North Sea Oil down from its 1999 peak by 45%, with depletion accelerating. The UK Department of Energy and Climate Change reported a 15.6% drop in the first quarter of 2011 against the previous year. Production from the Mexican Cantarell field, once the second largest in the world has in the words of one commentator "fallen off a cliff", dropping by nearly two thirds in five years. Saudi Arabia's Ghawar, triple the size of any other field and pumping steadily for half a century must now keep up its pressure by injecting sea water, which in some wells makes up to half of the output. See Matthew R. Simmons, *Twilight in the Desert*. NY: Wiley, 2006. Official figures for Ghawar are kept as secret as possible, since they are so critical to America's economic hegemony.

2.  Source: www.ranken-energy.com

3.  Many states in America have already given up on maintaining minor asphalted roads for cost reasons and are systematically returning them

to gravel, a step better than the "dirt roads" which once they were. About half of the 83 counties in Michigan, for example, have deliberately broken up little used tarmac roads by dedicated machinery, fragmenting and compressing the surface, which can be maintained simply by periodical smoothing by graders. The national situation has worsened since a landmark article in the *Wall St Journal* of 17/7/2010, titled "Roads to Ruin: Towns Rip Up the Pavement" with the strapline, "Asphalt is Replaced by Cheaper Gravel: 'Back to the Stone Age'."

4. cf. Richard Heinberg, Peak Everything. Gabriola Island, BC, Canada: New Society Publishers, 2010. William Stanton, *The Rapid Growth of Populations 1750-2000*. London: Multi-science Publishing, 2003; Albert Bartlett, *Arithmetic, Population and Energy*. DVD, University of Colorado, 2002. Stanton is the only notable writer to propose a solution, which calls for a degree of ruthless state control that Heinberg considers near-fascistic. The real debate has hardly yet begun. John Gossop's *Famine in the West* (Peak Food, 2007, and in e-book) is an admirable summary of the agricultural crisis now facing the world by a well informed farmer. See also David Ray Griffin, *Unprecedented: Can Civilization Survive the CO2 Crisis*. Atlanta GA: Clarity Press, 2014.

5. www.iea.doe.gov.

6. See David Mackay, *Sustainable Energy*. Cambridge: UIT, (also free online as PDF.) 2008. p. 9.

7. Peter Cox, Deepak Rughani, Peter Wadhams, David Wasdell, *Planet Earth, We Have a Problem*. 6/06/07. Available online as a PDF.

8. "Met Office report: global warming evidence is 'unmistakeable'", *Telegraph* online, 29/7/10.

9. James Hansen, *Storms of My Grandchildren: The Truth About the Coming Climate Change and Our Last Chance to Save Humanity*. London: Bloomsbury, 2009. p.6.

10. "Global Warming's Terrifying New Math," *Rolling Stone*. 19/07/12.

11. As quoted in "Warming will 'wipe out billions'," *The Scotsman*, 29/11/2000. In the face of such critically different estimates, those

who are in charge of the nation's energy policy must surely adopt the precautionary principle, but as the present work emphasizes, no one really is in charge.

12. Peter Christoff (ed.), *Four Degrees of Global Warming: Australia in a Hot World*. Routledge: NY and Abingdon, UK, 2013. p.4.

13. See Naomi Oreskes & Erik M Conway, *Merchants of Doubt: How a Handful of Scientists Obscured the Truth on Issues from Tobacco Smoke to Global Warming*. NY: Bloomsbury (2011). *The Press and Journal* (20/7/10) reported on the funding of climate sceptic organisations by Exxon Mobil, including the US-based front organisation Media Research Centre which was responsible for breaking the news on "Climategate". It also notes the refusal of Lord Lawson, a leading "climate change denier", to reveal the funding source of his Global Warming Policy Foundation. *The Independent* (24/01/13) reveals that the Donors Trust, which "funnels millions of dollars into the effort to cast doubt on climate change" through a third party *Knowledge and Progress Fund* is financed anonymously by Koch Industries, a large petrochemical company.

14. *Global Research* E-Newsletter, 10/7/10.

# Chapter 3

## Energy, Politics and War

*Geopolitics has been, to a great extent, synonymous with the politics of oil for five generations.*

Jeremy Rifkin, ***The Hydrogen Economy***

*The US maintains 737 military bases in 130 countries under cover of the "war on terror" to defend American economic interest, particularly access to oil .... The principal objective for the continued existence and expansion of NATO post-cold war is the encirclement of Russia and the pre-emption of China dominating access to oil and gas in the Caspian Sea and Middle East regions.*

Michael Meacher MP,
'The Era of Oil Wars', ***Guardian***. 29/05/2008

### 3.1 Oil and War

The scale on which energy as a commodity has shaped international politics and led to war is rarely appreciated, and raising awareness of this fact should be part of a strategy to promote microgeneration from renewables. One cannot begin to understand energy management without knowledge of the historical trajectory that the world embarked upon once oil was discovered and the internal combustion engine invented. The flection point when oil took over from coal can be dated with exactness to 1911, when Winston Churchill, then the First Lord of the Admiralty, made a momentous decision to change the Royal Navy from coal- to oil-burning, despite the fact that Britain had ample supplies of high grade coal but no oil. The words he used to explain his actions in later historical writings are as pointed a summary as one could find:

*The oil supplies of the world were in the hands of vast oil trusts under foreign control. To commit the navy irrevocably to oil was indeed to take arms against a sea of trouble …. If we overcame the difficulties and surmounted the risks, we should be able to raise the whole power and efficiency of the navy to a definitely higher level; better ships, better crews, higher economies, more intense forms of war power – in a word, mastery itself was the prize of the venture.*[1]

Oil has shaped international *Realpolitik* ever since, and the wars in Iraq and Afghanistan, Libya and Syria can be traced very directly to the struggle for "mastery" which began a century ago. As regards future oil wars, it is reasonable to assume that the drumbeat of American threats against Syria and Iran, which has vast reserves of high quality oil, are almost certainly the prelude to further conflict, delayed only until a new "coalition of the willing" has been assembled. America's sudden and surprise invitation to Russia, in December 2010, to cooperate with NATO can now be seen as a feint in a global struggle, made to neutralise a potential rival. From this perspective the so-called "War against Terror" – in which, at the time of writing, America now finds itself supporting Al Qaeeda in a Syrian civil war – can be seen as nothing more than a struggle for a dwindling global oil supply. This is the most obvious and least understood aspect of the global energy trap and America, which once had an over-abundance, has now become almost totally dependent on oil in other hands. The promise of a new bonanza of home-produced fracked shale oil and water flooding of old wells is at best only a temporary reprieve. The reality is, as Dick Cheney pointed out in a speech made in London in 1999, a year before he became the US vice-president, and effectively its CEO:

*By some estimates there will be an average of two per cent annual growth in global oil demand over the years ahead along with conservatively a three per cent natural decline in production from existing reserves …. Where is the oil going to come from? Governments and the national oil companies are obviously controlling about ninety per cent of the assets. Oil remains fundamentally a government business, While many regions of the world offer great oil opportunities, the Middle*

*East with two thirds of the world's oil and the lowest cost, is still where the prize ultimately lies.* [2]

In echoing Churchill's phrase "the great prize", Cheney is making the same statement about the necessity of going to war to obtain "mastery", but now war on a scale that can hardly be imagined. Were America to lose control over imported oil, it would be finished as a world power and would face economic collapse. Only when this simple fact is accepted can global politics be understood.

Every major political move by the US must be interpreted against a background of its energy vulnerability and the growing threat of its losing its monopolistic power to price oil in dollars. In a maze of confusing factors and the fog of real and economic war, these two factors are an Ariadne's thread. The incomprehensible situation in the Middle East makes dreadful sense when one accepts General Wesley Clark's assertion that US foreign policy has been based on taking out Iraq, Libya, Syria and then Iran in order to maintain American control of global energy. [3] With knowledge of these two factors, the K-gen and E-plus systems take on a new significance. As oil contains within it the seeds of war, they may be said to contain the seeds of peace. However, even on the most optimistic forecast, oil will continue to hold its top position for many years. Oil-based energy has become so profitable and so vital to the running of the modern state that it would be naïve to think that concern for the planet or the human family will be allowed to override shorter term interests of profit and convenience, or intellectual inertia. Oil has become the very oxygen of civilization, and a country deprived entirely of oil-based energy for only a week would collapse completely. Thus, however optimistic one may be about the potential for clean and renewable sources of energy, one must be realistic about time-scales and the magnitude of the challenges that a "renewables revolution" will bring with it, for the first priority is to keep society functioning.

To claim that the K-gen and E-Plus systems contain within them the seeds of peace may seem at first a fantasy of misplaced enthusiasm, but when one considers the way in which the modern world has been plunged into continual war by its need for energy and specifically for oil, the case for a comprehensive policy of renewable energy becomes insuperable. Since the connection between oil and war is not widely appreciated, it would be worthwhile to add a little more historical information to make the case.

The background to Churchill's action was the prior decision in the late 1800's by Kaiser Wilhelm II to build up a German navy to rival and surpass that of Britain. The Prussian empire was preparing for a showdown with the British empire and naval power was going to be absolutely critical. To have the advantage of faster and more efficient ships powered by oil (steam turbine or diesel) was a decision that made itself. There was no other option. Once made, however, it placed both nations at the mercy of oil supply, a supply chain which included reliable wells, pipelines, tankers and refineries. There was a further security problem, which Churchill stated to Parliament in 1913, "On no one country, on no one route and no one field must we be dependent." To achieve oil security he took two steps, the first being to introduce a bill that would enable the government to acquire a controlling 51 per cent of the Anglo-Persian oil company, later to be British Petroleum and then BP. Leaving aside the United States, Persia (today's Iran) was one of three giant oil sources, Mesopotamia and the Arabian peninsula being the other two. This fact set in train a reshaping of history and of the geography of the Middle East.

The second step in Churchill's strategy to establish secure supplies of oil was to cooperate with the other great powers of the day in setting up puppet regimes as far as possible in Persia, Mesopotamia and Arabia, and to do this the Ottoman empire, which held sway across the whole area, would have to be dismantled. The financial weakness of the Ottoman states, coupled with the onset of World War I, enabled this to be done, even though the allies' Gallipoli campaign, aimed at the capture of Constantinople, heart of the empire, was to fail badly. Two other important pieces in this complex political puzzle were control of the Suez canal and of Palestine.

When the political fragments had settled after World War I, the map of the Middle East had been completely redrawn. The Ottoman empire had effectively shrunk to Turkey and, authorized by the

League of Nations, the great European powers had seized the rest. France was awarded Syria and Lebanon, and Britain had obtained "the great prize", control of the oil fields, by acquiring Persia (Iran) and Iraq as "spheres of British influence" and Palestine as a "mandate", a new mutation of colonialism. Most of the Arabian peninsula had been handed over to the tribe of Saud, and with breathtaking chutzpah renamed Saudi-Arabia, all based on a tacit agreement that they would protect the oil interest of the Americans, and so it has continued to this day. Puppet kings were installed in smaller territories like Kuwait. The Suez canal had been made politically safe through a hundred year lease to Britain,

This arrangement endured more or less until the end of World War II, and it is worth recording that Hitler's gamble on attacking Soviet Russia in 1941 was largely forced on him by Germany's need for a secure oil supply from the fields of Baku and Ploesti. Likewise, the Japanese attack on Pearl Harbor in the same year was a direct result of its need to escape the stranglehold on its oil supply that America had been exercising. No sovereign country can survive in the modern age without adequate energy in the form of oil, let alone a country at war. The geopolitics of oil started to change radically towards the end of the Cold War. America's home produced oil, once seemingly inexhaustible, started to run down rapidly due to profligate use, Iranian oil became less secure after the Shah (a British and American puppet) was deposed, the oil and gas potential of the Caspian Basin started to be opened up, and the energy demands of a rapidly industrializing China became a major factor. With very limited oil resources of its own, China could easily find itself in the desperate position of Japan in 1939, ever in danger of political blackmail by those countries which controlled the oil.

These are some of the most basic facts in a global kaleidoscope of national energy needs. The whole scene is too complex to summarize adequately, but there is ample information in books and articles. Perhaps the most comprehensive is Daniel Yergin's Pulitzer Prize winning, *The Prize: The Epic Quest for Oil, Money and Power* [4] and it is noteworthy that he took his title from the same phrase used by Churchill and Cheney in the quoted statements above. Less directly addressing the theme of oil and war is Geoffrey Sampson's *The Seven Sisters: The Great Oil Companies and the World they Shaped*, first published in 1975 but recently reissued [5] and still providing a good historical background. Zbigniew Brzezinski's *The Grand Chessboard: American Primacy and its Geostrategic Imperatives* [6] is valuable in showing how the epicentre of oil production has shifted eastwards and is recreating the "Great Game" of world control.

## 3.2   The Great Game and World War III

The term "Great Game", popularized by Rudyard Kipling, originally referred to the struggle for dominance between the Russian and British empires where their interests clashed in the Indian subcontinent. It was, however, the use of the term by Halford Mackinder that has kept it alive and given it a new meaning and a new threat. For Mackinder the Great Game was the struggle for world dominance through military control of the Eurasian heartland. What has changed since his seminal paper, "The Geographical Pivot of History" (delivered in 1904 to the Royal Geographical Society) is that the Great Game has become the struggle for oil and natural gas. That fact is significant in the present context of micro- and mesogeneration for two reasons. Firstly, as earlier noted, it explains the wars in Iraq, Afghanistan and Syria and America's economic war with Iran. Secondly, it highlights the fact that so long as nations compete for oil energy, the danger of war will be real and present and since the relevant nations now are continent-sized, war when it comes will surely be global and nuclear. New alliances are being formed, as they were before the First World War. The North Atlantic Treaty Organization is under the effective control of America, though nominally headed by the Danish Eurocrat and uber-hawk Anders Rasmussen, at the time of writing, and now has treaties or agreements with such north Atlantic countries as Australia and Mongolia. The Shanghai Cooperation Organization, founded by Russia and China in 2001, ostensibly for trade and friendship, makes pointed reference to the fact that it "speaks for half of humanity", implicitly the half that has not aligned itself with the US/NATO global ambition.

It hardly needs to be explained today that the invasion of Iraq by the Americans and British had nothing to do with Saddam Hussein possessing weapons of mass destruction, but everything to do with control of Iraqi oil, which was about to slip out of America's grasp, as Saddam had announced his intention of limiting payment for oil to euros. Iran too was all set to go the same way, which would have had a devastating effect on the value of the American dollar. The war in Afghanistan, while very much in line with Mackinder's theorizing, has been justified as a necessary step in a global war on terrorism, but the reality is that control of Afghanistan was deemed necessary to control the pipelines that were being constructed to transport oil and natural gas from the Caspian basin. After the "humanitarian bombing" of Libya to establish freedom and democracy, an interim government was put in place (including members of Al Qaeeda) and their first pronouncement was that the new Libya would be governed under Sharia law. All this is, however, secondary to the main aim, for Libya's oil resources are now under new management.

Far more than fossil fuel energy is at stake, for the infrastructure that America has built up since the Soviet Union retreated from Afghanistan, "the new Silk Road," as it has been called,[7] is a transit route for military equipment as well as oil. As the Indian analyst and former diplomat M. K. Bhadrakumar puts it:

> *The resuscitation of the Silk Road project to construct an oil and gas pipeline connecting Turkmenistan, Afghanistan and Pakistan (the TAPI pipeline) will need to be seen as much more than a template of regional cooperation .... TAPI is the finished product of the US invasion of Afghanistan. It consolidates NATO's political and military presence in the strategic high plateau that overlooks Russia, Iran, India, Pakistan and China [and] provides a perfect setting for the alliance's future projection of military power for crisis management in Central Asia.*[8]

For "crisis management" read "future energy wars".

The brute fact is that while every civilized nation needs oil and gas, we are destroying civilization as well as the climate in our desperate need to claim our share. Only a few countries have enough oil within their geographical boundaries to be self-sufficient in energy, and the rest of the world therefore is either in conflict or making alliances in preparation for conflict in a struggle to survive. Few have been as clear-sighted about the global politics of oil and gas as the late and indefatigable campaigner Hermann Scheer, who used his position as a member of the German parliament to raise awareness of it and whose words may be used here as a summary:

> *The only reliable means of avoiding the global resource conflict is to go cold turkey on fossil energy as soon as possible. Instead, however, the transport networks for fossil resources are being expanded to allow larger and faster flows, and countries are beginning to arm for the coming conflict.*[9]

How prophetic his words were has been shown by recently declassified documents obtained by the American public interest group Judicial Watch, which detail the rise of ISIS from a US funded and armed grouping of Islamic terrorists. What began as an attempt to overthrow the legitimate Syrian government in order to clear the way for a pipeline to bring Qatari gas to Europe got totally out of hand, but the most surprising fact to emerge is that the US government (or that branch which now determines foreign policy) deliberately and consciously took the risk in order to neutralise Russian and Iranian energy interest in the region. In the words of the investigative reporter, Nafeez Ahmed, "The Pentagon foresaw the likely rise of the 'Islamic State' as a direct consequence of this strategy, but described the outcome as a strategic opportunity to 'isolate the Syrian regime'."[10] Before ISIS suddenly burst upon the world in 2014 as the resurrection of a Sunni Moslem caliphate, fully armed and funded, it was regarded by America, and one must presume with tacit agreement of the UK and other western countries, as a strategic asset in the global energy war. The genie is now out of the bottle and in a hugely complex situation, with many "unknown unknowns," it is hard to see how resolution can be found other than the default strategy of all out war.

The threat of war has been increased by Saudi Arabia's unilateral decision at year-end 2014 to force down the world price of oil to little more than half its value six months previously. The geopolitical reasons for this are far from transparent and the consequences are unpredictable. One effect is, however, already clear: the bipolar world of the USA and Soviet Russia, which gave the world forty years of a kind of peace in the form of the cold war, has now changed to America + NATO versus a Russian-Chinese defensive alliance, putting aggressive pressure on Russia's borders, along with sanctions that amount to economic war. This re-alignment does not bode well.

The continuing threat of a third world war would effectively disappear if every country were reasonably self-sufficient in energy. Implementation of the K-gen and E-plus concepts would certainly not solve this at a stroke, but if it was seriously attempted, they would together go a very long way to decreasing the present geopolitical tensions. Vigorously promoted and generously funded by government in their early stages of development, they promise a real alternative to a future where "ignorant armies clash by night" in wars for control of the planet's fossil energy supplies. Matthew Arnold's phrase from his famous poem *Dover Beach* expresses in a nutshell the tribal world's natural response to conflict of interest. Can the growing competition for the world's dwindling oil supply be resolved not by "ignorant armies" but by engineering skill and logistical teamwork of a high order? That is the challenge now facing a world on the brink.

### 3.3   The Game Changer

The global energy situation has changed in many ways since the first draft of the book was completed. International awareness was raised at the Global Warming Conference held in Paris in December 2015, although its conclusions did not go much beyond declarations of intent. There have been dramatic developments in the field of battery-driven and hybrid cars, but they remain outside the budget of most people. Uptake of photo¬voltaic panels in the UK has slowed drastically, since the government slashed subsidies, and many installation companies have gone out of business. So too in Germany, where promotion of photovoltaic systems has effectively ceased, as the barrier has been hit that sooner or later will confront any expansionary programme, when the Grid will have far more energy than it can use (and has to pay for) in the summer but far too little in the winter. Without a revolution in energy storage, such as the book will propose, this dilemma can only be resolved by retaining all the conventional generating capacity as back-up, regardless of cost.

Without doubt the greatest change in the global energy picture has come from the reduction in the price of oil from $120 to $40 a barrel, as a consequence of Saudi-Arabia's unilateral decision in 2015 to flood the market with oil. There are complex political and economic reasons for this unexpected move but, oversimply, it has been triggered by unforeseen advances in drilling technology, which have enabled previously inaccessible oil and gas to be recovered, resulting in a situation where the US has effectively achieved self-sufficiency. Figures (at August, 2016) from the US Energy Information Administration show only 24% of the country's needs being imported, but a comparable amount exported and a huge surplus of natural gas. The Saudi government had hoped to kill off competition from America and Russia and, in effect become a monopoly supplier, but its strategy seems to be failing. The longer term global effects are not easily predictable but the effect of very cheap oil on global warming is, alas, all too predictable, for if oil remains around $40 a barrel or even less for a sustained period, there will be a drastic loss of momentum in the renewable energy industry. In the last analysis people and nations will buy their energy from the cheapest provider, whatever fine sentiments are expressed and applauded at international conferences.

### References

1.  Winston Churchill, *The World Crisis* (1923), Vol. 1, p. 134.

2.  Quoted in F. William Engdahl, *Full Spectrum Dominance*. Wiesbaden: edition.engdahl, 2009. p. 56.

3.  Wesley Clark is the former Supreme Allied Commander Europe of NATO. In a 2007 speech he records his surprise at discovering a "policy coup" in the US administration which was directed at "removing governments in Iraq, Syria, Lebanon, Libya, Somalia, Sudan and

Iran." The speech has been widely reported and can be seen on You-Tube. Search "Gen. Wesley Clark – Exposes US Foreign Policy Coup."

4. Daniel Yergin, *The Prize: The Epic Quest for Oil, Money and Power*. NY: Simon and Schuster, 2009 [1991]

5. Geoffrey Sampson, *The Seven Sisters: The Great Oil Companies and the World they Shaped*. London: PFD, 2009 [1975].

6. Zbigniew Brzezinski, *The Grand Chessboard: American Primacy and its Geostrategic Imperatives*. NY: Basic Books, 1998. It is understandable that Brzezinski would use the Great Game as a quasi-permanent frame of policy into which he makes America's global interests fit, sometimes awkwardly. As a Pole, he is acutely sensitive to the expansionist instincts of Tsarist and Soviet Russia.

7. Not to be confused with the Chinese initiative, with Russian, Indian and other countries' cooperation, that has also been called the New Silk Road. This is a quite awe-inspiring vision of a high speed rail, road and shipping system, already under construction, centred on the manufacturing hub of Zhejiang province, with a horizontal axis running from Vladivostok to Madrid and various north-south axes connecting to St Petersburg and Siberia, Mumbai, Bangalore, etc. Among other effects, it will create a new Eurasian commercial unity, drastically reduce container ship traffic and cut travel times by anything up to 80%. See, e.g., Pepe Escobar, "China's Pivot Toward Europe May Cut U.S Out of Deal," www.financialsense.com. 21/12/14.

8. M. K. Bhadrakumar, "NATO weaves South Asian web," *Asia Times*. 23/12/10.

9. Hermann Scheer, *The Solar Economy: Renewable Future*. London: Earthscan, 2002. p. 105. Scheer died in 2010 of unexplained causes of a vague internal nature, while beginning a lecture tour in America. The official cause of death, without post mortem, was heart failure, but the alternative press regarded him as another victim in the dirty war for control of global energy, "heart-attacked" as David Kelly had been "suicided" seven years earlier.

10. Nafeez Ahmed, "Pentagon report predicted West's support for Islamist rebels would create ISIS, in the online magazine *Insurge Intelligence*, 22/5/2015.

# Chapter 4

# The UK Crisis

*In Europe we are facing the risk of the lights going off. This is not a joke.*

Fatih Birol, Chief Economist, International Energy Agency

*Britain is sleep-walking into an energy crisis that threatens Seventies-style black-outs and huge increases in bills.*

Headline summary of Ofgem Report,
"London Evening Standard," 3/2/10

## 4.1   When the Lights go out in London

The 2009 Ofgem report on the present and future state of Britain's energy needs, entitled *Project Discovery*, was described by the business editor of *The Times* as a "disaster scenario between the lines." The crisis that is foreseen arises from the fact that fully a third of the UK's generating capacity is due to be taken out over the next ten years: eight major oil and coal-powered stations are to be decommissioned by 2015, and seven of the nation's ten nuclear stations by 2020, resulting in "involuntary curtailment of demand" for some large energy users. These dates can be extended in emergency by stretching various safety factors, but not indefinitely, as the rundown schedule has already commenced in some instances. The sober fact is, as one former executive of the National Grid points out, is that "virtually all coal-fired stations [are] in excess of thirty five years old, having been constructed for a design life of thirty years."[1] Furthermore, as regards the coal-fired stations, any extension would potentially have to be agreed by a change to European law, which was passed with a view to lowering carbon emissions. As a member of the EU, the UK is now legally obliged to generate 32% of its electricity from renewable resources by 2020.

The UK government's decision to go for nuclear was announced by Tony Blair in February 2005, and after years of vacillation, the government has committed itself (in May 2012) to a program of nuclear generation, by offering a guaranteed profit to the contracting companies, French and Chinese. Such a guarantee is in effect a moral hazard, tempting the company or companies involved to pad their accounts in order to maximize this bounty. From the point of view of free market enterprise, this solution is the worst of all possible worlds, and the open-ended cost will be paid for by the helpless consumer. At February 2015 EDF, the French energy giant, is still negotiating terms with the UK government and discussing areas of responsibility with its Chinese partner. Even without delays (for which such projects are notorious) the first new nuclear station can hardly come on stream until 2022 by the most optimistic forecast, and in the words of Alistair Buchanan, the chief executive of Ofgem, "the scale of collapse in energy supply between 2013 and 2017 is profound and worrying ... and the really sweaty-palm moment in terms of possible shortages." Those who are old enough to remember the miners' strike in early 1972 will understand the significance of the crisis that Buchanan is foreseeing. It resulted in ten power stations being shut down, random disconnection of electricity supply for up to nine hours daily and, on the worst days, two million people being laid off work and thousands of electric trains being cancelled. On a personal note, I recall travelling by rail one February evening across a darkened country to arrive at a mainline station lit only by a few candles at the ticket barrier. That gives a very brief illustration of what is at issue.

The future supply situation merits the term "nightmare". The Institution of Civil Engineers has long warned of the danger, and has projected a shortfall in 2016 of 40% of the current peak load. [2] The historian and political commentator Max Hastings writes in the strongest terms, "If you do not believe me about the gravity of the threat, read the energy report published two months ago by the Royal Academy of Engineering, This presents a terrifying picture of government inertia and folly." [3] Current hopes that the crisis can be avoided by replacing the UK's energy needs with wind-generated electricity are illusory. The illusion is widespread and is partly explainable by assuming the existence of a commercial lobby seeking only its own profits, and partly by a wilful ignoring of the obvious fact that wind-generated electricity will always require an equal standby capacity. These vitally important issues are looked at more closely in chapter 5, *The Great Wind Farm Non-solution*. The reality is exemplified by the quite typical situation of the weekend of Feb 12th, 2009, during which temperatures were low and energy demand high, wind-generated electricity in the UK dropped to 0.4% of national need, and coal-generated electricity demand increased from 35% to 50%.[4]  The fact is that in a more severe and prolonged cold and windless spell, every existing generating station would need to be called upon, and therefore it would be folly to close any of them down in hopes that wind farms could eventually replace them. True, while the wind is blowing, less coal will be required to fuel the national grid supply, thus ensuring less carbon pollution, but it is by no means sure that the grid will be flexible enough even to take full advantage of this potential benefit. Neither coal nor nuclear powered stations can be switched on and off at will as the wind rises and falls.

This raises a wider problem about the National Grid, namely that it is configured to distribute very high voltage electricity from a few large power stations that were built, for obvious reasons, close to large coal mines that are now largely closed. The existing trunk and branch pattern is thus poorly fitted for either macrogeneration relying on offshore wind farms or from a diffuse national feeder system of millions of microgeneration units. In its present form, it is highly wasteful of energy, as shown earlier in Fig. 1.

### 4.2   The Political Response

For a variety of political and economic reasons, the government has had no coordinated national energy policy for thirty years. Before the advent of privatisation in the early 1980's, the Central Electricity Generating Board (CEGB) had both coordinating and executive authority, but subsequent bodies set up by both Conservative and Labour governments function essentially as advisory committees to

Parliament. In theory such a structure should work to create a national policy, but in practice it is impossibly slow and compromised by the overarching need to maintain some semblance of competition between the electricity companies in a pretend free market. Moreover, these initially British companies are now transnationals, some headquartered in France and Germany, and their prime concern is not the welfare of the British community but profit.

Britain's looming energy crisis could have been seen ten years ago, but few of the politicians that we elect to look after the nation's interests seemed able to understand its urgency. The 2009 Ofgem report, cited above, on the country's future energy supply future, was entitled *Project Discovery*, but virtually repeats the same facts given by the 2002 report of the Performance and Innovation Unit commissioned by Tony Blair. This was debated in Parliament but neutralized by optimistic spin. The lack of any sense of crisis or of a plan to meet it was evident then in the response of the Environment Secretary Margaret Beckett. She simply redefined the central issue not as a crisis but as a stroke of commercial good luck, saying that, "Energy efficiency and renewables will certainly have a key role to play and will offer great opportunities for innovative businesses in the UK." The only serious initiatives are the lumbering dash for nuclear mentioned above, which has been added on top of the decision by the then Energy Secretary Ed Milliband in July 2009 to build 7,000 more wind turbines and the scheme of domestic grants and feed-in tariffs to encourage microgeneration. The latter, now being drastically cut back, actually originated not as a government initiative but as a private member's bill tabled by the then Conservative MP, Peter Ainsworth. His vision and persistence deserve recognition, for his bill was designed to incentivize, in his words, "a fundamental shift in the way Britain produces its energy - away from the model of centralized power that ruins the environment to a model that makes homes and business less dependent on foreign fossil fuels."

The elephant in the room in *Project Discovery*, is the fact that government intervention is now needed directly and on the largest scale to solve this national energy problem, and this would be a reversal of the policies of the Conservative and New Labour parties, both of which are driven, often blindly, by ideological conviction that a maximum of commercial activity should be left to market forces but, ironically, heavily subsidized by the taxpayer. The profit motive is without doubt a great incentiviser, but it runs head on here into the needs of the community. Some individuals with logistical vision can see this clearly. Ofgem's chief executive says that leaving the privatised system to supply future needs is "not an option," and the consultancy McKinnon and Clarke spells out the reality in starker terms as the possible need for "Soviet-style central planning." Discounting the dramatic phrasing, the truth is that a national energy policy cannot be cobbled together from the present fragmented structures, driven by profit and otherwise unconcerned with society's needs. Faced with the logical certainty of an energy shortfall, the response of both political parties in government has largely been akin to that of a rabbit trapped in a car's headlights. Completely absent is recognition of the reality that without action as urgent and coordinated as in a war situation there will be some ten years when the country's energy needs cannot be met. Britain cannot survive as an industrial nation for so long a period on, at best guess, seventy per cent of its current electricity supply. This is why microgeneration, long considered a frill, must be considered a main strategy. If promoted single-mindedly, it could be adding to national energy supply within two or three years and providing significant new capacity just as withdrawal of current supply is becoming critical.

The National Grid has warned that the risk of electricity shortage is at its highest level for seven years and the situation is "likely to get worse before it gets better" with the safety margin in the event of a cold winter dropping to 5%, half the level of 2012. Chris Train, marketing director, told an industry conference that "things will be tighter than they have ever been historically," but continued in optimistic vein, "While there have been power station closures since last winter, the information suggests that the market can meet demand in cold weather. But as the system operator, we're never complacent and it's up to us to be ready to balance the system in real time." [5] He calmed public apprehension by saying that in the event of potential shortfalls it would be industrial users, not the voting public, that would be the first to face restrictions of supply, but gave no indication as to when, or how, this worsening situation would or could improve. It is cold comfort, in a very literal sense, to know that the National Grid is not complacent about future electricity supply, which is not within their control anyway, despite his reassuring words.

## References

1. Derek Birkett, *When Will The Lights Go Out? : Britain's Looming Energy Crisis.* London: Stacey International, 2010. p. 193.

2. Institution of Civil Engineers. *Energy Policy: Key Issues for Consultation*, 8/12/03. On the Internet as a PDF.

3. Max Hastings, "Why I fear the lights will go out." *Mail Online*, 24/5/10.

4. Source: *The Times*, 16/1/09.

5. "Electricity shortage raises risk of rationing," *i Paper*. 8/10/13.

# Chapter 5

## The Great Wind Farm Non-solution

*Were these wind farms truly efficient and capable of resolving our power needs, I might be persuaded to grit my teeth, and endure their ugly intrusion, but in fact they are almost useless as a source of energy.*
James Lovelock, ***The Vanishing Face of Gaia***

### 5.1  Costs, financial and environmental

The initial generosity of the feed-in tariffs offered by the government, but now drastically reduced, clearly indicates that the Department of Energy's commitment to microgeneration is uncertain. Its commitment to macrogeneration through large scale wind turbine technology, however, remains unquestioned, for there has been no cutback in its vast investment programme during the same period. Strictly, the commitment is to windmills, which work on a quite different principle to turbines, but while that is a very important engineering distinction, it can be overlooked here, for the significant fact is that the government has made electricity generation from wind farms a main plank in its energy policy. The intrusiveness to which James Lovelock refers in the words quoted at the head of this chapter may be gauged from the fact that wind farms of upward of a hundred units are to be located on high ground in many areas of natural beauty, and the size of the units merits the term "monster". A typical wind turbine stands more than half as high as Blackpool Tower, is topped by a nacelle the size of a single-decker bus and has a blade sweep half the area of a football pitch. The swooshing noise they collectively make can carry well over half a mile, and the subwoofer "thump" is strong enough to be picked up by a seismometer. When one considers how strict is the legislation that prevents one from putting up an advertising billboard, the way in which these environmental monstrosities go through planning procedures almost on the nod, is hard to fathom.

The financial cost of these windmills is equally massive, and the hidden costs could double the tax burden on future taxpayers. In rough figures, the "catalogue price" of an onshore windmill rated at 3 megawatts is close to £1 million and an offshore probably closer to £1.5 million, when the additional costs of ships, floating cranes and underwater cables are added in, but these figures do not include several factors which conceal the true long term costs. The following items indicate some of them.

Firstly, the total capacity is distorted by measuring it in terms of households that it can notionally supply, although this is easier for the ordinary person to understand than kilowatts or megawatts. Thus the Siemens company can claim for its wind farm near Glasgow, built for Scottish Power in 2006, that it will produce "enough electricity for 200,000 households," but the reality is that for long periods it will produce only enough to supply a tenth of that number. To quote an industry expert:

> *The aggregate power of the wind farm will vary hundreds of times during a year, between full capacity of 322 MW and near zero, [which] means that Scottish Power has to keep about 290 MW from conventional power stations available for speedy backup when the full capacity of 322 MW drops sharply with subsiding wind speed in order to prevent a serious blackout that could spread over a large part of England. A consequence of this is that one or two conventional power stations must be kept running at reduced power, and therefore with reduced efficiency. This means that more $CO_2$ will be exhausted per produced kWh.* [1]

Secondly, even in the best of conditions each unit will produce less than half the "nameplate" or rated capacity and nothing at all for long periods. The government's strategy document assumes a load factor (average compared to peak or rated capacity) at 42% for off-shore, but current data shows a maximum of 29%. [2]

Thirdly, to the cost of the generating equipment must be added other factors which are not immediately apparent, not only the infrastructure, which involves building approach roads onshore and new high voltage transmission lines, but in modifying the Grid to enable it to take large and unpredictable new capacity from the wind farms as they are constructed. The E.ON *Wind Report 2005* has estimated that this has cost about 3 billion euros in Germany to accommodate about 9,000 on-shore wind turbines. Ofgem has estimated that to implement the UK wind farm programme will entail additional infrastructure costing £20 billion, roughly the total value of the existing grid.

Fourthly, perhaps the most overlooked cost factor is maintenance and replacement, for the strains on such giant installations are extreme, with a single turbine blade weighing ten tonnes or more. Maintenance is a particular problem for offshore wind farms, because of salt water corrosion, which calls for more expensive design in the first instance. Denmark is the acknowledged leader in offshore wind generation, and the experience of its *Horns Rev* field off the west coast is instructive. One analyst reports, "All eighty turbines experienced an almost complete breakdown as a result of the penetration of saltwater spray. (The build-up of salt on the blade ... has been shown to reduce the generated power by 20-30%.) Each turbine had to be completely dismantled ....Everything was transported from that location at sea back to the factory for repair and for design changes. A programme aired on Dutch television on November 4, 2004, showed only four turbines were turning, one was 'temporarily out of service', and of the remaining 75 only the useless decapitated pylons were standing. It was a grim sight." [3] Lessons have certainly been learned from these early failures, but the problem of maintenance is rarely aired publicly and long-term replacement almost never.

The root problem of wind generated electricity on every scale is unpredictability but, as will be explained later, coordinating and smoothing of the demand/supply curves can be more easily achieved at the level of microgeneration by short and long term storage. At macro-level the problem is acute, because wind energy increases exponentially, as the cube of velocity. Thus, up to Beaufort scale 4 (moderate breeze) a mere 4% of rated capacity is generated, at scale 5 (fresh breeze) it is only 20%, and only at scale 6 (strong breeze) is a useful 43% generated. Then it leaps to 100% at Beaufort 7 (near gale) and beyond that the blades must be feathered and the wind turbine taken out of service. The energy source has to be turned off just as it peaks, and attention given not to harnessing it, but to designing and integrating dump load technology.

The same source notes that if each installation is checked every three months, there will need to be 320 visits per year, and "how these could be and would be carried out is a mystery. How many helicopters and specialized maintenance ships with very high lifting facilities would be essential for this maintenance, to say nothing of the number of personnel?" The end result is that "Denmark needs to reckon with an on-call supply from either Sweden or Germany … on an irregular basis [which] will of course be considerably more expensive than the kWh produced in a traditional manner in Denmark itself. It is clear why Denmark has the highest electricity tariffs in Europe." [4]

Such information is never aired in the optimistic forecasts of the UK, and other governments, nor is the fact that wind turbines actually consume electricity, particularly in keeping them facing the wind, Despite all these negatives, the Department of Energy presses on with plans to better the EU target of 20% and produce up to 32% of electricity from renewable sources by 2020, almost all from wind farms. This calls for questioning. How can so many energy experts be so wrong? How can the UK, and many other governments, commit themselves to a renewable energy policy which is so expensive, so non-productive and so destructive of the environment? A complete answer to this question is urgently needed, but apart from the propaganda effect of emphasizing the word "green" and dismissing honest concerns as selfish "nimbyism", the deliberate distortion of facts can be traced to two so-called authoritative documents on which government advisers and policy-makers have had to rely. Another, and more worrying, possibility emerges when one studies the role played by incentives, and seeks to account for the enormous subsidies which are being paid to the small number of companies which manufacture, install and maintain the equipment.

The fanfare of publicity which accompanied the opening of the Thanet wind farm in September 2010 makes no mention of the true cost which includes a gift from the British taxpayer to Vattenfall, the Swedish owners, of £1.2 billion in subsidies over the estimated twenty years of the turbines' working life. This sum invested in a single nuclear power station would yield thirteen times more electricity and with much more reliability. For Vattenfall this represents a secure annual return on capital of 13%.[5] When the former Minister for Energy, Chris Huhne, boasted that this is "only the beginning," one must wonder what green fantasy land our political representatives inhabit. Only a fortnight later Ofgem published the figures about the real world, and they are cause for the greatest alarm. To accommodate the wind farm programme, the national grid will have to be effectively rewired, costing the astonishing figure of £200 billion over the next decade. To give the full picture it must be said that this will include an upgrading amounting almost to a whole new infrastructure that would have had to be made in any case to modernize the National Grid. This would enable developments in micro- and mesogeneration generation to be integrated, so in this respect the kickstart given by wind farms may prove to be a blessing in disguise. Ofgem estimates that upgrading the Grid will cost the average householder £6 per year, on top of an average annual electricity and gas bill of £1,200. One suspects that this may be a tactical understatement, but it would be a price well worth paying.[6]

We may be grateful to the Renewable Energy Foundation, a green think tank, for probing into the real costs of "free" wind power and the government's dysfunctional strategy. Typical of the facts they have unearthed, which receive little publicity, are the huge sums paid to wind farms to switch off their energy when the Grid does not have the capacity to use it. In April 2011, for instance, the National Grid paid out £900,000, twenty times the value of the power that Scottish wind farms were capable of producing, in order to persuade them to switch it off. In theory, this electricity could have been sent south to England, but the transmission lines did not have the capacity to handle it. Looking at the overall picture, a typical wind turbine generates electricity worth about £150,000 per year and attracts a subsidy of £250,000.[7] As campaigner Struan Stevenson, Scottish MEP, aptly put it, "They are not farming wind but farming subsidies." Good business for someone, but the profit ultimately comes from the taxpayer's purse, via their utility bills. At present wind farms produce a notional 5% of the country's requirements. It needs only simple arithmetic to work out how much we shall all be paying for this "free" energy when, and if, Britain meets its target of producing 20% of its energy needs from renewables in 2020. Already in 2013 taxpayers, through their electricity bills, are paying energy firms £30 million to turn off their windmills, often dumping in effect a half of their production. The figure is sometimes higher then this in windy conditions. Onshore farms receive in subsidies about double the wholesale price of electricity, and offshore about three times.[8] One wonders at what point someone in government will point out the obvious, that this is economic lunacy.

There are huge sums at stake here, and as with the hushed-up bribery scandal that accompanied BAE contracts with Saudi Arabia for fighter planes, it would be naïve to assume that all is above board in the matter of contracts for wind farms. The former Labour MP Alan Simpson, speaking at the Solar Century Event, strongly implied the existence of an energy cartel: "Current energy policy in the UK is dominated by the vested interested of 'Big Power'," adding significantly, "The national grid is monumentally inefficient as an energy system. It was a half-decent idea for the middle of the last century, but 70%-80% of energy put into the grid disappears before you or I even switch the light on. We need not an energy, but a power revolution that takes control from the centre and literally puts power back into the hands of the people." There is, indeed, the strongest circumstantial evidence of corruption, when it comes to wind farms. The Siemens company, by far the world's largest supplier of wind turbines, was fined a record £523 million by the US authorities and another £354 million in Germany after being found guilty of bribery in other areas of their business. [9]

## 5.2   Finding the Truth

The two main sources of information on which the UK government's *Renewable Energy Strategy Document* relies are The British Wind Energy Association (BWEA) and the University of Oxford's

Environmental Change Institute's (ECI) reports, more particularly, *Wind Power and the UK Wind Resource* (2005). Both of these sources are suspect, but for different reasons. The BWEA can be expected to put the best gloss on figures, since the profits of its members depends on making the most convincing case to governments, and there is a thin line between glossing and falsifying the figures. The subsidies themselves, based on Renewable Obligation Certificates (ROCS) and fines for non-compliance, are of Byzantine complexity, involving legislation that forces utilities to take electricity from the primary generators at a high cost and pays them to do so. The additional cost, on top of the wholesale price, has been estimated at £60 per mWh. [10] To put that figure into perspective, every time a householder does an hour's ironing or cooks a meal they are silently taking about 3p from the taxpayer and putting it in the pocket of some large company.

Although the ECI is not a profit-making organization, its figures, which are the basis of UK government policy, are so completely at odds with the performance data from functioning enterprises, such as the Danish government's and the German energy company E.ON's *Wind Report 2005*, that some explanation is needed to account for the anomaly. As one obvious example, the compiler of the ECI 2005 report, Dr Graham Sinden (now on the Renewable Energy Advisory Board) argues that "with a good dispersion of wind turbines, the variability of wind output over the UK as a whole can be expected to be smoother than output from any individual site or region" and "low wind speed conditions affecting 90 per cent or more of the UK would occur in around one hour every five years during winter." This is simply not true, or true only if the word "low" has hidden qualifications. The German experience, as quantified in the E.ON report, states (on the basis of 7,000 dispersed turbines), "On Christmas Eve 2004 wind production in Germany fell 4,000 MW in ten hours, representing the capacity of eight large coal-fired plants" and in the typical October of 2007, there were 19 days when average wind across Germany was only 8% or less of rated capacity, including some days at zero. The Danish experience, as reported in the magazine *Politiken*, is that 84% of wind-generated electricity had to be exported – i.e. dumped at cheapest prices – and wind turbines were often shut off because their output did not match peaking demand.[11] This snapshot clearly illustrates an aspect of the wastefulness of wind macrogeneration that is rarely noted. It is possible that the situation may be improved in the future by linking the wind turbines with electromechanical storage (flywheels) on the same scale, but that would be a whole new initiative with comparable costs and there has been no evidence so far that it is part of the plans of either the government, the utilities or the manufacturers and installers.

Rarely noted is the ecological devastation that is involved in the production of the giant magnets which are at the heart of the wind turbines. The largest of them requires a permanent magnet made of neodymium-iron alloy weighing over two tons. Without notable exception the neodymium is mined and smelted in China, creating air, water and ground pollution on a scale that can truly be called nightmarish. One of the main factory complexes, at Baotou, pours all its acidic effluent and seven million tons of toxic tailings into a lake which has grown to be five miles across and which, according to a recent report in a UK newspaper, "instantly assaults your senses …. your eyes water and a powerful acrid stench fills your lungs." One can imagine the permanent effect on those who lived in the vicinity. Greenpeace China says, "There is not one step of the process that is not disastrous for the environment …. Villagers living near the lake face horrendous health risks from the carcinogenic and radioactive waste."[12] This is the hidden human cost of the wind farms, to add to a commercial and governmental confidence trick that is stealing from every householder. The same article puts figures to this: "Already, the Renewable Obligation legislation adds £1.4 billion to our bills each year to provide a pot of money to pay power companies for their 'green' electricity. By 2020 the figure will have risen to somewhere between £5 billion and £10 billion."

When all the complex facts have been assessed, the simple truth is that because wind-generated electricity is so unreliable, and no way of storing very large amounts has yet been seriously attempted, for every megawatt of electricity that wind farms produce, at enormous cost to the taxpayer and the loss of amenity value to the countryside, another megawatt of electricity must be added to the national capacity from conventional gas or coal fired power stations. Thus the "green credentials" of wind farms are a fraud. Far from reducing carbon dioxide emissions, every wind farm adds more, since the standby facilities must be kept running at reduced power and thus with reduced efficiency. Germany has spent £4 billion on wind farms in the past ten years but not a single conventional power station has closed.

This is the Monty Python world of energy into which the country is being led through the simplistic thinking of some obsessive green activists, the limited vision of politicians and the lobbying power of vast commercial interests. These three influential streams have come together to ensure that the public is being systematically lulled into supporting the wind farm programme through an unthinking belief that "wind is good, because wind is free" and "wind is good because wind is green." The truth is that we are sleepwalking into an energy disaster as we watch our countryside despoiled. While other renewable sources are neglected for lack of funding, our political representatives in Westminster press on with what has been called "Government sponsored robbery of the poor for the benefit of the rich."[13]

### 5.3   The Promise of Wind Power

Despite all the negative arguments put forward in this chapter, the inescapable reality is that wind is a powerful, clean and limitless source of energy. Harnessing it is a challenge to our imagination, and the inescapable problem of energy storage which goes with intermittent wind supply is being tackled by inventions on the small and large scale. The emphasis in the book is on microgeneration, but several proposals on the large scale call for mention. The first, which has attracted substantial investment from Bill Gates, has been called a "drop dead simple energy idea" and "gravel on ski lifts"[14] and is an extension of the concept of pumped water storage, as used when wind-generated electricity is married to hydroelectric. Energy Cache Inc. has a functioning prototype in California, which uses surplus electricity to raise huge buckets of gravel to a height of about 100 metres, and discharge them when there is demand for electricity. The falling gravel drives an electrical generator just as falling water does in hydroelectric installations. It is claimed to cost only 40% of hydro but one must wonder how much maintenance and replacement of machinery will be needed to cope with the enormous wear and tear on such a system. A different solution, but on meso, rather than macro-scale, is the Vanadium Redox Flow battery, an Australian invention which exploits the unique chemical properties of vanadium and a special electrolytic fluid to create a rapid recharge function. Typically, such a battery (or an array) will be rated at about 300 kilowatts and give about six hours supply. This is useful in an emergency or load-levelling situation but does not address the basic problem of wind-generated electricity, namely, that one can quite easily have several days of calm weather, when ten times this amount of back-up capacity is called for.

There are also new developments in chemical engineering which store wind-generated energy in the form of molecular bonds of liquid, non-polluting fuels. This process hinges on using nitrogen extracted from the atmosphere as the feedstock for ammonia and also for nitrate fertiliser. The process is energy intensive, but now instead of using fossil fuel to generate the required input, one can store the intermittent energy of the wind by using it to create ammonia from the unlimited and free source of air. More will be said on this in the Appendix, in the section, "The Promise of Ammonia". Suffice here to say that this can in principle reduce the vast wastefulness of monster wind turbine farms. It does not, however, remove their insult to eye and ear, nor the damage they inflict on the quality of life and house values of those who live near them.

The K-gen system comes at the problem from the other direction, seeking to harness wind energy on the microscale, using electromechanical batteries – i.e., flywheels – for storage. This will be treated in Chapters 6 and 9. The K-gen system will also incorporate new ideas for increasing the collection area of wind turbines and also extend existing designs that make use of the roof of the house as a wind collector and exploit magnetic levitation bearings, which are effectively friction-free. This enables the energy of low wind speeds to be captured, when ordinary turbines do not turn at all.

## References

1.  William Hyde and John Webley, "When the wind stops – the other side of the wind turbine argument." Article prepared by Kentish Weald Action Group. 23/7/2008. Available on the Internet. Hyde is a Chartered Engineer and retired South Eastern Electricity Engineering Manager.

2.  Hyde and Webley, *op. cit.*, p.5.

3.  J. A. Halkema, "Wind Energy: Facts and Fiction" (2004). Available on the Internet.

4.  Halkema, *op cit.* p.38.

5.  "The prices of ensuring the lights don't go out", *The Independent*, 5/10/10.

6.  The figures are from Christopher Booker, "The Thanet Wind Farm will Milk us of Billions," *Telegraph online.* 25/9/10.

7.  "Wind farms paid £900,000 to switch off," *Sunday Times.* 1/05/2011.

8.  Nick Modermott, "Windfarms paid £30 million a year to stand idle because the grid can't cope with all the energy they produce." *Mail Online.* 9/08/13.

9.  "Record US fine ends Siemens bribery scandal," *The Guardian*, 16/12/2008.

10. John Etherington, *The Wind Farm Scam*, London: Stacey, 2009. p.79. This is an invaluable source book and develops the arguments in this chapter much more thoroughly. It should be noted that the author gives a plausible and detailed argument against the connection between atmospheric CO2 and global warming. However, his judgement is made without reference to contrary data that has appeared since 2009.

11. Eric Rosenbloom, "A Problem with Wind Power." www.aweo.org/windbackup. This paper is a compact and invaluable source of hard data.

12. Simon Parry and Ed Douglas, "In China, the true cost of Britain's clean, green wind power experiment: Pollution on a disastrous scale." *Mail Online.* 29/1/11

13. Hyde and Webley, *op. cit.*, p. 5

14. Katie Fehrenbacher, "The story of Energy Cache, a drop-dead simple energy idea. *Gigaom.com/cleantech/the-startup-behind-bill-gates-ski-lift-for-energy-storage.* Retrieved 27/3/12.

# Part Two

# Fundamental Principles

*We have to evolve means for obtaining energy from stores which are forever inexhaustible, to perfect methods which do not imply consumption and waste of any material whatsoever. I now feel that the realization of that idea is not far off.*

Nikola Tesla (1897)

*Discovery consists of seeing what everybody has seen and thinking what nobody has thought.*

Albert Szent-Gyorgyi

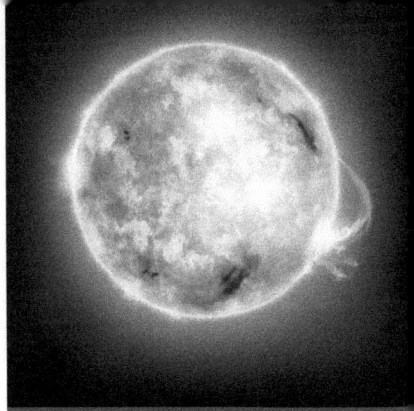

# Chapter 6

## Theoretical Foundations

*The Engineer is a mediator between the philosopher and the working mechanic .... Hence the absolute necessity of possessing both practical and theoretical knowledge.*

Henry Robinson Palmer, founder member of the
Institution of Civil Engineers (1818)

### 6.1  E-plus in Concept

Although the E-plus house and the K-gen generating system can be seen as a single system, there are advantages in treating them separately, since a central function of the E-plus is energy conservation, whereas the K-gen, is concerned almost entirely with energy generation. This way the K-gen system can be approached simply as an engineering challenge, and this will be done in the following chapters. The E-plus house will then be seen more clearly as a radically new architectural concept and will be treated this way in Chapter 10. A particular advantage in focusing microgeneration on the single dwelling is that research and development will be driven and financed by the market, thus avoiding the vagaries of government policy and all the problems inseparable from financing very large projects. Once an economically efficient prototype house is in existence, it may be expected to attract research funding from commercial sources, and thus avoid the endless and frustrating problem of seeking grant aid.

The term "E-plus", short for "Energy-plus", is used as an umbrella term to indicate a type of house which sets out to maximize its potential for generating electricity and at the same time minimize its energy requirements.  It is not to be confused with the low-energy or zero-carbon or autonomous house, though it shares important characteristics with them. It aims to be

- more than self-sufficient in energy production
- the essential unit in a larger system providing baseload electricity
- a planned step to making the UK self-sufficient in energy
- a key element in a strategy to reverse global atmospheric pollution.

These are very ambitious aims, but that does not in itself make them unattainable.

In practice there will have to be two quite different kinds of E-plus house, for the challenge of integrating the most efficient ideal system in a new-build house becomes an untidier matter of compromise and trade-off when retrofitting an existing house. While the former might be expected ideally to generate 200% or more of its energy requirements in the best conditions, a much lower performance is to be expected from the latter, often less than 100%. These estimates must be no more than best guesses until the system has been designed and one or more prototypes built. There are about 180,000 new houses built annually in the UK, against an existing housing stock of about 25 million, with annual demolitions of about 18,000. From these figures it is easy to see that the real challenge will lie in retrofitting existing houses, but there is every reason to start with a blank sheet of paper and design an ideal house, modifying and compromising thereafter as may be required.

The idea of many millions of houses providing baseload electricity is not entirely new. Greenpeace UK issued a paper in 2005 urging the need for decentralising and downsizing of electrical power generation in what it foresaw as an energy internet in which "buildings, instead of being passive consumers of energy, would become power stations, constituent parts of local energy networks, where coal-fired power stations have been closed and their surrounding webs of pylons dismantled, restoring swathes of countryside …."[1] Few commentators on energy are able to think so radically, but George Monbiot's account of how the logic of microgeneration came home to him is worth quoting:

*I have been looking at the problem the wrong way round. I have been thinking about electricity and heating fuel as almost everyone has been thinking about them since construction of the national grid began in 1926: as commodities supplied over great distances from major sources. But there is an entirely different way of responding to the question of how our energy might be generated. It's generally described as "micro generation" or "the energy internet". In its pure form, it involves scrapping the national grid. Instead of producing a large amount of power in a few places, the energy internet produces small amounts of power everywhere.* [2]

The E-plus vision, if it may be so called, differs from this only in assuming that the national grid should remain, albeit in modified form, for the foreseeable future. The important thing at this stage, however, is the need to raise awareness of the national and global energy problem that is bearing down on us, and the potential of microgeneration to solve it.

It would be easy to dismiss microgeneration out of hand as totally impractical, without examining and assessing the case for it. Certainly it will call for a revolution in thinking about energy management and in architectural design as well as engineering, but initial cause for optimism may be taken from the miniaturisation in computers that has happened in the last thirty years. Until 1980, when Hewlett Packard launched its first PC, computers were bulky fixed installations, often taking up a whole room and the idea of a personal computer small enough to fit on a desk would have been considered science fiction. Nevertheless, imagination, research and investment made it happen. More than this, in the last ten years the computer world has been totally rearranged: the mainframe, which previously was "the market", is becoming almost a niche, the desktop revolution has become history, and all the excitement and the main source of profit is now in the field of tablets and smart phones, which are really hand-held computers with functions that until twenty years ago could hardly have been imagined outside science fiction comic books. In less than ten years the smart phone has gone from concept to reality and is now being manufactured in hundreds of millions. It is changing society in unpredictable ways and, quite literally, changing human consciousness.

By analogy with the evolution of the computer, the future of electrical generation can be envisaged as a process in which meso-generation, analogous to the PC, replaces centralized macrogeneration, and microgeneration, analogous to the laptop, tablet and smart phone, replaces both as the focus of the market. Macrogeneration covers centralized power stations using fossil or nuclear fuels, wind farms, large hydro installations in some countries and, exceptionally, tidal and river barrages. Mesogeneration, or distributed generation, would supply a factory or small community and microgeneration would normally supply the needs of a single house. There is not always, however, a clear dividing line between meso- and micro-, since there is no difference in practice between making a single house or ten houses or a factory self-sufficient in energy. What sets the E-plus/K-gen system apart is that it aims to work in conjunction with the Grid to supply energy when needed and smooth the national supply and demand curves. This is the greatest challenge that any national grid must face, and it needs no specialised knowledge to realise how important and how difficult this can be. In any 24 hour period the national requirement will go from a maximum demand, when everyone is at work, to a fraction of that when they are asleep. It is a subject for mild humour that at half time in an international football match, demand for electricity spikes, as millions of people switch on their electric kettles to make a cup of tea. It is, however, a most serious business for the control engineers, who can only meet this surge in demand by holding several large central installations on standby and ramping them up well in advance to cope with an extreme demand that may last only fifteen minutes.

Seamless integration of microgeneration with the Grid will have to come as a second stage of development, but the fact that it is anticipated makes the K-gen concept more complex than anything yet proposed in this field. Meeting the challenge calls for a system integrated both vertically and horizontally, rather than simply built up from unrelated bolt-on devices that either generate or save energy. Viewed vertically, it must work as part of a centralized Grid system but one which in the future will be layered in a different way, fed from many medium scale systems producing electricity from renewable sources.

Viewed, horizontally, the E-plus and K-gen systems together must link at least the following functions into a coordinated whole:

- solar energy collection and conversion
- heat pump technology
- wind-powered generation
- energy storage
- insulation
- low temperature central heating

### 6.2   K-gen: A Metasystem

The K-gen concept can be defined through five operations which it performs on energy. These may be listed mnemonically with words all conveniently beginning with the letter C, namely:

- Collection
- Conversion
- Concentration
- Conservation
- Control

Each of these operations may be regarded for research purposes as a system in its own right, and the K-gen system, bringing all together and, as it were, orchestrating them, may thus be regarded as a metasystem. The significance of these operations and an initial assessment of their feasibility will appear as each is examined separately. This will bring out the engineering challenge, which will be dealt with in Part Three.

### 6.21   Collection

The feasibility of both the E-plus and K-gen concepts must begin with an estimate of how much energy from the sun is available. In a 1900 article "The Problems of Increasing Human Energy" Nikola Tesla estimated that the sun heats the earth with the equivalent of four million horsepower (about 3m kilowatts) per square mile, though it obviously varies between the equator and the poles, and this represents a quite astonishing amount of potential energy to be harvested. Tesla's broad figure can be made more meaningful by scaling it down to domestic size in a UK latitude. There is a significant difference in the amount of solar energy available as one goes from south to north. Eastbourne, for example, receives about 1,850 hours of sunshine per year, Birmingham 1,350, Glasgow 1,250. Also, the quality of the light – i.e., its energy content – decreases as one travels north and the angle of incidence diminishes. The term "average house" may be hard to define but is taken here to indicate broadly a three-bed semi, with a footprint of about 80 sq. metres and a garden of about 150 sq. metres and situated around 52 degrees of latitude, i.e., with about the same amount and strength of sunshine as Birmingham. What is most efficient here may not work so well in other parts of the country or the world. In a continental climate, which covers most of North America and Europe east of, say, Germany, there will be much more sunshine in winter but much lower temperatures to cope with, while in summer there will often be need for air-conditioning. Nevertheless, it is a reasonable conjecture that once the principles have been identified, the practice may be adjusted to maximize electrical generation in other climatic conditions.

Over the course of a year the house and garden would receive about 1100 hours of usable sunshine, but much more per day at the height of summer than in the depth of winter. Taking the spring and autumn solstices as a mean, each square metre would receive about 3 hours of sunshine per day at a rate of about 300 watts per hour, or somewhat less than 1kWh of energy per day. In December this would be reduced to perhaps 0.2 kWh and in June increased to perhaps 2 kWh. The exact figures are not important at this stage, the point being to get some idea of magnitudes as a prelude to estimating the number of photovoltaic or thermal panels required to optimize solar energy collection year round. Two square metres should, theoretically, be more than required in summer, but far from adequate in winter. Following this line of reasoning, if a tenth of the total available area (roof + garden) were to be used, leaving the garden more or less intact, there would be space for 23 square metres of panel, far more than the householder's requirements. The question then is what proportion of this can be converted into usable electricity?

The efficiency of solar panels is variable, and has been steadily increasing for many years through research and fierce competition. Photovoltaic panels ten years ago averaged 6-8 per cent conversion from solar energy to direct current electricity. In 2013, the norm is about 12-15%. Solar thermal panels are much more efficient, with the recently invented vacuum tube technology capable of over 80% efficiency in converting solar energy to hot water and, moreover, they perform much more efficiently than PV panels in hazy conditions and especially in winter. In summer solar thermal panels will often produce far more hot water than is needed, and for that reason a control system needs to be incorporated, either to dump or store the excess. The most efficient use of energy produced by solar thermal panels is in the direct production of hot water for domestic and heating purposes, but the K-gen system also aims to turn hot water into electricity. Between the two systems, producing part electricity and part hot water, a rather arbitrary conversion figure of 33% can be put in, purely as a temporary marker. That is to say, overall a third of the sun's energy would be potentially convertible into electricity or electricity equivalent. Using this and the estimated insolation figure above, one can make a provisional and very wet finger assessment of the potential input of energy from the sun at point of use as approximately 4,000 kWh per year.

This figure must be set against an annual household energy consumption of about 20,000 kWh,[3] and at first may seem to leave an unbridgeable gap. However, when consumption is broken down into categories, a quite different picture emerges, for no less than 12,000 kWh of household energy per year is used for space heating and domestic hot water, and since the Passivhaus model has shown how the heating requirement can be reduced to less than 1,000 kWh with disciplined attention to insulation, we are left with a much more achievable target figure of about 9,000 kWh, before exploring other renewable sources, such as wind energy and heat pumps.

Domestic lighting consumes about 600 kWh, and although LED technology is still in its infancy, it already has the proven potential to reduce the electricity requirement for lighting the average house by 80-90%, thus closing the gap by another 400 kWh. We thus end up with an annual demand of about 8,500 kWh, of which solar energy can provide about a half, and if a little more of the garden were to be used for freestanding solar panels (photovoltaic or thermal) this figure could be raised to three quarters.

From these broad brush figures, it can be seen that a preliminary case at least can be made for the proposition that energy from the sun alone, if maximized, could satisfy most of the energy needs of a domestic residence. However, the figures must be treated carefully since, as noted, the most efficient component produces hot water, not electricity. A good heat pump would multiply the heat energy from the hot water by four or five, but there is clearly a limit to the amount of hot water a house needs, for heating, washing and cleaning, so the engineering challenge now shifts to how to convert the low-grade energy obtained from the sun in the form of hot water into usable electricity. This will call for critical innovation, and some fairly sophisticated engineering, much of which has already been achieved, but is scattered around in specialist journals and on the Internet. A major function of the book is to bring together some of this information.

Converting the heat energy of water into electricity and storing large amounts of either heat or electricity until required are the most obvious challenges of the K-gen system. First, however, the problem needs to be defined and quantified, and at the present time there does not appear to be even the basic data of how many kilowatthours of energy the average home would need to store through the winter in order to be self-sufficient.

Wind collection presents a quite different range of challenges and possibilities. It is generally assumed today that efficiency of wind turbine technology is limited by the Betz law, which sets a limit of efficiency at 59.3 percent While this is generally true, it is rarely noted that the law itself makes two assumptions that the K-gen system would call into question, namely that the collection area and back pressure of the wind turbine are limiting constants. The assumptions are valid with windmills and conventional wind turbines in general, but there is no reason why the collection area cannot be enhanced or the back pressure artificially reduced, thus in principle increasing efficiency. More will be said later on this, for it is a clear engineering challenge.

Collection of ambient heat from air or ground source is a function performed by the heat pump. While heat pumps may be said loosely to collect heat energy from subsoil or outside air (and are usually marketed in these terms) their function is primarily to concentrate low grade heat to a useful density, using electricity from the Grid to do so. Thus they consume electrical energy and provide heat energy. Understanding the theory behind them opens up a new understanding of energy collection. Strictly, a heat pump functions to create the conditions for state change, and then captures the energy released in the transition from gaseous to liquid or solid state. It is the fact that the trigger energy required is less than the energy released that puts the principle of the heat pump at the heart of the E-plus/K-gen system.

Another and largely neglected aspect of energy collection is waste heat recovery, which is in large part an architectural problem and may be left for later treatment.

## 6.22 Conversion
Each of the above collection systems must be coupled with a conversion system. PV panels convert sunlight into direct current and this must be converted to alternating current at an acceptable voltage for feed-in to the Grid. This is achieved by an inverter, which typically uses 6% or 7% of the input energy in the process. Improvements in inverter design are ongoing and there is every reason to anticipate some reduction in this figure when research has run its natural course, thus placing less demand upon the National Grid when the use of PV panels has become more general. If this higher efficiency can be achieved and PV panels installed in a million homes, one or even more power stations could be closed down. This fact is worth noting, since it illustrates the leverage that can be exercised by what

may seem at first very small incremental gains in efficiency at the level of the individual house.

Wind energy must be converted to rotary motion and torque in order to generate electricity but begins as air pressure exerted on turbine blades. A simple propeller connected through a gear train to an alternator is the intuitive way to harness the wind, but there are many complexities concealed in such an arrangement which will suggest more efficient ways of converting air in motion to electricity. An understanding of how air in motion actually behaves is an obvious, but often neglected factor, calling for knowledge of fluid mechanics and considerable mathematics and wind tunnel testing. Then there are the friction losses of the gear train, which the design engineer must seek to minimize. At another remove, the use of permanent magnets in the standard design is both costly and puts a future limit on expansion because of the world shortage of rare earths, especially neodymium, which quadrupled in price between 2009 and 2011. A supply squeeze by China, their main source, could virtually bring the use of high efficiency permanent magnets to an end. Although such estimates must be very approximate, it is thought that the planet's resources of rare earths will be effectively exhausted in less than thirty years. It is logical therefore to plan now for that eventuality and explore other technologies. Of those which come to mind, the AC induction motor is perhaps the most obvious, as it is already being developed by some auto manufacturers to counter future exhaustion of rare earth elements.[4]

## 6.23 Concentration

Concentration of solar and wind energy is essentially compression in time and/or space and is necessary to create and store energy density sufficient for the system to do the work of generating electricity. Solar energy is densified by using mirrors or lenses and, in a different kind of situation, light energy is concentrated to perform work by using laser technology to compress photons in space. Concentration of wind energy is a curiously neglected area in energy management, and the K-gen system focuses on increasing the collecting area of the wind and using the extra energy in a process of self-compression to dramatically increase the efficiency of the turbine. Little more can be said here other than generalities, as patent applications are in process.

### 6.24 Conservation

There are three aspects of energy conservation, the most obvious requiring no technology at all but merely using energy efficiently and not wasting it through draughty doors or windows or by leaving lights switched on when not needed. There is more than enough general literature on this freely available from various sources, such as the Energy Trust, and nothing more need be said here.

The same literature deals with the second aspect of conservation, thermal barrier technology, which is called for in two critical areas, namely, to retain heat (or sometimes cold in warm climates) within the four walls of the house and to make long term heat storage possible. Structural insulation is a well developed part of architectural technology, but there are very recent technological advances which have not yet been incorporated into mainstream practice, notably aerogel, which has been called "a vacuum flask on a roll". These will call for closer examination later.

The third aspect of conservation concerns methods of storage of energy in the form of heat, electricity, chemical or mechanical-gravitational batteries, phase-change materials, hydraulic accumulators and supercapacitors. This is a vast but fairly well defined area, presenting both tested information and research projects to be thought through and developed. There is a significant overlap with fuel cell technology, though the fuel cell is strictly a through-flow energy device, rather than a means of long term storage. Even a brief review of the full range of options is well beyond the scope of the present work, and calls for a serious research project. The preferred solution for a K-gen system will be electromechanical, or flywheel, storage, which is well developed in public transport, racing car and UPS (uninterruptible power supply) applications. Without doubt technology can be borrowed from all three fields. Once electromechanical storage on the small and medium scale becomes central to energy planning, a radically new perspective opens up, for it can be seen that in principle the flywheel can function much as hydroelectric storage but with the great advantage that micro- and mesogeneration do not require expensive transmission lines and so avoid subsequent transmission losses. All depends on whether or not flywheel technology has the kind of capability that is needed for long term storage, and new developments strongly suggest that it has.

### 6.25  Control

The function of a control system is to regulate and link together all the component parts of the system but also to link it to the Grid. There are therefore two functions, and coordinating them to function as one presents a serious challenge. While its immediate purpose is to maintain a balance between the energy demands of the house and the available supply, a control system must also find a balance between the demand and supply requirements of the Grid. As this is recognized, the key role played by storage in the whole system (Grid plus micro/meso-installation) comes into focus. The individual house or cluster of houses will be a negligible item in the concerns of the Grid for some time to come. Yet if the K-gen system, perhaps in modified form, lives up to expectations and is widely adopted, it will eventually need to be integrated with the Grid's national strategy. The fact that none of this exists at present should not be taken as proof that it can never exist. Once renewable energy can be harnessed in sufficient quantity and stored, there would be no reason for continuing to burn fossil fuels to generate electricity.

The nature of a control system now becomes more apparent. There must be a collection system which can provide an overall surplus of energy from the houses to the Grid on a year-round basis and the Grid (which itself has no storage facility) must be able to regulate the amount of electricity that is drawn from the multipoint storage system. This line of thinking leads to an unexpected conclusion, that a major requirement of the control system must eventually be how to deal with excess capacity. This is a problem inherent in solar thermal technology, which produces in the summer far more hot water than is needed by the individual house. It is probably fair to say that no one has thought about how to deal with an excess of electricity, since the overwhelming problem has been how to generate enough of it. However, if the K-gen system, which aims to convert most of the heat from solar thermal panels into electricity, is proven and widely adopted, an answer will have to be found.

## 6.3   The Very Smart Grid

National Grid is a publicly quoted company, one of the world's largest utilities, owning some 4,500 miles of transmission lines and 340 substations. It has always been a smart system, if the term is taken to mean a complex system which must incorporate feedback mechanisms and information flow to function at all. Its overarching challenge has been to match national supply and demand, thus creating in effect a steady state. To maintain that state there are critical choices of several kinds to be made, most obviously in predicting demand but also finding an economic balance. At its simplest this involves drawing upon the most expensive producers of electricity only as a last resort in order to meet peak requirements and disconnecting them first as demand falls. There are also subtler considerations involved, such as the economics of running several power stations at less than capacity or switching off two or three altogether and restarting as demand builds up. This has always been a complex operation and the need to integrate wind turbine generated electricity, which comes intermittently and often in violent bursts has not only made things vastly more complex, but reduced the overall efficiency of the Grid. Adding micro- or mesogenerated electricity to the current situation can only make it more complex still and, in fact, will call for rethinking the form and function of the Grid. Already there are enough photovoltaic panels producing electricity to have stimulated new thinking about smart grids across the world, for the output from a PV panel rises from zero to a peak every day before falling back to zero and to a very much flatter curve in winter.

The shape of the future may perhaps be seen in the Flexicity company, a private venture which was set up to coordinate alternative electricity supplies from businesses which either generate their own electricity or have standby generators. It is an ambitious project and is now well established as an intermediary, enabling electricity to be taken from many small suppliers and fed into the Grid to meet peak or emergency demand. Flexicity itself is not involved in the electrical generation business but operates the control and switching system that is at the heart of the operation, so that many small individual sources can be mobilized to meet need as it arises and arrange for payment to them, with all functions automated once they have installed the initial equipment required by the suppliers. These are without notable exception businesses with their own diesel powered generators, and while they add to atmospheric carbon dioxide, they produce far less than the coal and gas-fired mainline generators which they supplement briefly in periods of high demand and which otherwise would need to be in kept in "hot standby" mode. The creation of Flexicity is a heartening development for microgeneration of all kinds, since the conceptual and actual structure created by the company can be transposed without change to handle K-gen systems on any scale.

Though it is never in the headlines, a great deal of thinking about smart grids is being done. One might, indeed, talk of very smart grids, insofar as they incorporate a central control to switch supply on an off to individual houses, and even to specific appliances, in order to keep the whole system in balance. A recent technical publication *Smart Grid: Infrastructure and Networking*, with contributions from many countries, goes so far as to assert that smart grids could be "the biggest technological revolution since the invention of the Internet" and that "we cannot even imagine the products and service that will evolve as the smart grid takes hold."[5] Since one can only manage what can be measured and sensed, these two functions underlie all the rest, and one contributor's definition of smart grid is worth quoting at length to show how comprehensive these functions are:

> *A smart grid generates and distributes electricity more effectively, economically, securely and sustainably. It integrates innovative tools and technologies, products and services, from generation, transmission and distribution all the way to customer devices and equipment using advanced sensing, communication and control technologies. It enables a*

*two-way exchange with customers, providing greater information and
choice, power export capability, demand participation and enhanced
energy efficiency. [6]*

The smart grid is intelligent in a way that the conventional grid does
not aspire to be. It is in the words of another contributor, adaptive,
predictive, integrated, interactive, optimized and secure.

Despite the government's financial incentives, micro/mesogen-
eration is still a negligible part of the Grid's supply and, understand-
ably, of its future plans. However, if small scale generation increases
to a significant extent - say, ten per cent of demand - the energy
management problem will become more clearly defined and, thus,
more amenable to the sort of innovative logistical treatment just
mentioned. The core problem is always how to match complex
supply and demand curves, mostly day against night and summer
against winter. Supply from wind sources is year round but erratic
and unpredictable, while supply from solar sources is at its maximum in summer but demand is at its greatest during the winter.
There are also differing demands for electricity between weekday
and weekend. Typically, consumption nationally drops by almost
20% from Friday to Saturday.

As regards the economic factor from the Grid's perspective, simply bringing more large installations on line as demand increases
is the most expensive and least efficient. In decreasing order of efficiency, hydro, gas, oil and coal fired installations each require time
to reach top output from a cold start, from half a minute to several
hours. Looking far into the future (one assumes) and a fully developed system of microgeneration from renewables, the existing Grid
structure could be turned on its head, with the shortfall from a nationwide microgeneration system being made good by having a very
small number of coal- or gas-fired generation stations whose capacity could be increased as required. However, as the following chapters will explain, this standby function can be in theory fulfilled by
using state of the art flywheel technology. Flywheel energy storage
has so far has been limited almost entirely to creating uninterrupt-
able power supply systems (UPS) for the use of hospitals and vital

IT installations. There is every reason to think that with relatively small modifications the principles which underlie UPS systems can be applied to resolving the problem of energy storage which dogs all renewable energy systems.

This chapter has attempted to give an eagle's eye view of the E-plus and K-gen as concepts as a first indication of the potential as energy-generating and energy-conserving systems. Later chapters will give a more detailed picture of the engineering and architectural challenges and first suggestions for a research programme required to proceed to proof of concept.

## References

1. Greenpeace UK, "Decentralizing Power: An Energy Revolution for the 21st Century", 2005.

2. George Monbiot, *Heat: How to Stop the Planet Burning*. London: Penguin, 2006. p. 124.

3. This figure and others are taken from the *Carbon Independent* website.

4. cf. "Nikola Tesla's Revenge," *The Economist*, June 2, 2011. Strapline: "The car industry's efforts to reduce its dependence on rare-earth elements has prompted a revival in the fortunes of an old-fashioned sort of electric motor."

5. Krzysztof (Kris) Iniewski, (ed.), *Smart Grid: Infrastructure and Networking*. NY: McGraw Hill, 2013. p. xv

6. *Ibid.* p. 7

# Chapter 7

# K-gen and the Philosophy of Engineering

*The products of engineering, and even the way that engineers work, can offer enlightenment on some of the enduring questions of philosophy.*

Natasha McCarthy, *Engineering: A Beginner's Guide* (2009)

## 7.1   Engineering as a Social Force

The K-gen system embraces complexities at various levels and represents a breakthrough in the philosophy of engineering. Formerly a non-subject, the need for a philosophy of engineering is slowly attaining recognition. At a 2008 workshop organized by the Royal Academy of Engineering papers were delivered with such titles as "Maintenance as Morality", "Beyond the Modern Profession: Rethinking Engineering and Sustainability" and, most intriguingly, "The Engineer's Identity Crisis: *Homo faber* vs. *Homo sapiens*". [1] The E-plus and K-gen concepts, it needs hardly be said, are vitally concerned with sustainability and the moral imperative of caring for a planet. Together they may well prove to be a seed for change in the engineer's identity. That there is need for a new social identity is not widely recognized and will not be argued here, but it is worth noting that whereas science and engineering were once regarded as a prime means for building civilization, their alliance with Big Business and the global armaments industry and particularly with the fossil fuel industry has slowly but critically changed this unspoken attitude. The inherent social idealism of engineering has not only been lost to view but technology is now often associated with a suspicion of what might be called Frankenstein science that would have once been unthinkable.

Through the five component K-gen concept – as revolutionary in its own way as was the idea of a steam engine three centuries ago – a global society corrupted by so-called financial engineering has the chance now to return to a more innocent and inspiring age, when the engineer was recognized as a visionary in action, playing an essential part in enriching society and building civilization. We are being pushed to write a new chapter in the romance of engineering by the urgent demands of the global energy trap. Society once had a vision of a world transformed by the steam engine; now we need a vision of a world transformed by being able to harness unlimited clean energy. We know it is there: the engineer-scientist's task is to make it happen.

## 7.2   K-gen and Systems Thinking

The most obvious connection between philosophy and engineering lies in the use of systems thinking, which emerges naturally from engineering practice. The deepest significance of systems thinking and operational research lies, however, in the fact that they represent a self-consciously new way of looking at problems. To this extent a truly modern philosophy should take cognizance of systems analysis, although academic philosophers trained largely in linguistic analysis may find the assertion questionable. Properly applied, a systems approach can open up logical horizons, enable dots to be seen, as it were, and then joined up, problems to be better defined, assessed in multiple contexts and seen as inter-related subsystems. The general value of systems modelling in clarifying thought before "plugging in the figures" may be indicated in figures 2 and 3 below, which also illustrate the different mindsets at work in dealing with fossil fuel and renewable energy generation.

Before coming to that, it is perhaps worth noting that any small items in the house that diminish energy needs can be considered loosely as part of the E-plus system, even though they do not actually generate electricity and are not themselves systems, since they will affect the figure of merit, that is, the benchmark against which its efficiency is measured. Energy saving at domestic level extends to identifying such things as eco-kettles, economy shower heads, dehumidifiers, voltage optimizers, etc., and particularly low energy light bulbs, which will call for special attention in the chapter on the architectural challenge. The planner starts without knowing in advance how much energy such items can save, or even how many are on the market, or what inconvenience might come from striving for maximum economy. Hence from a team research point of view it would seem logical to put all these potential items in one box initially, consider it a composite system along with the others listed below and hand it over to a task force to come up with all the information and recommendations required when theory moves on to proof of concept. This "miscellaneous" box can be dropped quite naturally into the systems model that will be suggested below.

The K-gen system is a revolution in energy management in several ways, which will emerge as its different aspects are explored. It is also a new concept of engine, as will be explained in Chapter 11. Early theory and practice arose from making the most efficient use of coal in driving the steam engine, in effect releasing the energy of its molecular bonds and reducing hydrocarbon to ash through exothermic reaction. Subsequent developments in internal combustion engines worked on the same basic assumption, that energy is obtained by destroying the chemical bonds of liquid and gaseous hydrocarbons. In both cases it is understandable that research focused on extracting a maximum of energy stored within the fuel, and minimizing waste heat. By contrast, the modern approach, in its quest for sustainable and clean energy, is to harvest unlimited renewable energy and then manipulate it to produce electricity. Since renewable energy is not diminished or disordered in the process, the key concept of entropy in thermodynamics, and heat engineering in general, is irrelevant. The K-gen system strives to minimize not entropic but parasitic loss: that is to say, loss of heat or electricity in the processes of collection, conversion, transmission and storage within the system.

The different approaches are shown in the following process diagrams, the first illustrating how conventionally energy is generated by release of chemical bonds through heating until the bonds break and the second by creative manipulation of natural energy.

**Triggering Energy**

**Energy Source & Storage**

| match | → | tinder | → | kindling | → | coal |

**Energy Conversion**

| work | ← | torque | ← | expansion | ← | steam |

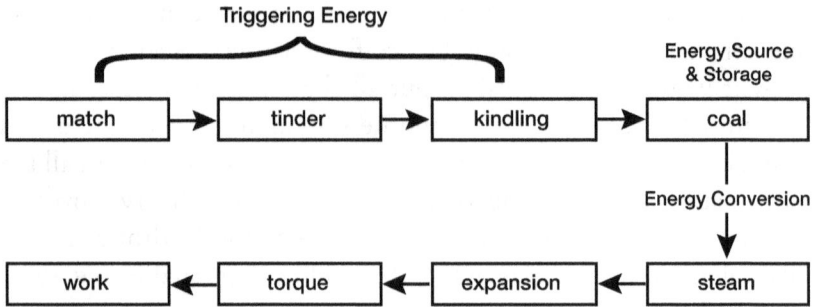

Fig. 2  Fossil fuel energy flow (steam engine)

The first three stages may seem at first rather whimsical, but make a very important point, namely that *heat energy is needed to release the energy stored within the fuel.* This is true of both the external (coal-fuelled) and internal (petrol- or diesel-fuelled) combustion engine. Anyone who is old enough to remember cranking a car to start it on a cold morning will need no convincing of this. When Watt was building steam engines, it would be normal to start one up from cold by striking a spark from a flint, which would ignite tinder, which would ignite kindling, then firewood, the heat from which was needed to release the energy of the coal.

The K-gen/E-plus schematic below is clearly more complex than a fossil fuel system, and as the nature of its complexity is understood, deeper insight is gained into the challenges, which become more clearly defined and thus more capable of being overcome. Of these, two will come to be seen as most critical to the feasibility of the project, namely the need to convert the energy of hot water into electricity and to store surplus energy until required.

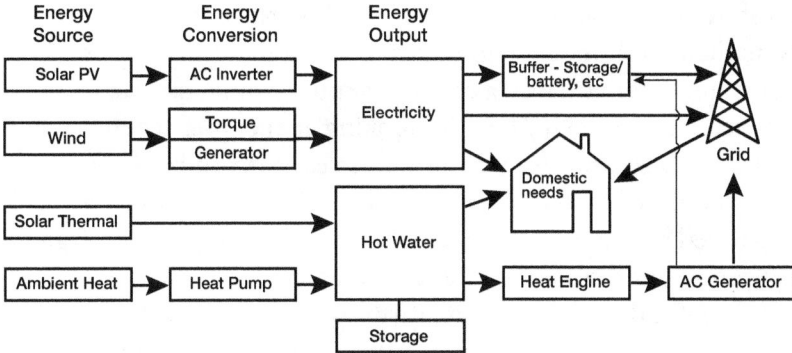

Fig. 3   K-gen renewable energy flow

As regards the most appropriate systems model, little more can usefully be said here. The most general model, System Dynamics, based on the three components of flow, stock and feedback loops, has some obvious appeal, with solar and wind energy being regarded as flow, the storage component as stock and a control system as feedback. It is more useful to take the basic principle of homeostasis from physiology, for viewed from this standpoint it can be seen that while the Grid is a system seeking homeostasis (dynamic balance), the owner of the E-plus house or commercial installation will be seeking profit from maximizing energy output. It is by no means impossible that at some future time, there will be a national surplus of electricity from microgeneration and this will come into conflict with the requirements of the Grid. Indeed, this situation has already arisen in Germany.[2] At the present time in the UK the Grid is in an impossible position, charged with implementing a national policy but existing as a PLC with a prime function of maximizing shareholder value and only restrained in its monopoly position by the regulatory body Ofgem, which makes arbitrary decisions that are often political compromises.

### 7.3   Learning from Nature

The engineer used to approaching energy generation in terms of fuel-burning will need to reorientate himself or herself to see the harnessing of renewable sources in terms of energy release, energy transformation and balance of forces. The principle of entropy and its mathematical modelling, which have been the keystone of thermodynamics for well over a century is counterproductive in this new situation. In the changed way of looking at energy as renewable and cyclical, nature provides various examples that will be of help. Perhaps the most obvious is the hydrological cycle, in which water is evaporated by the sun on the earth's surface and the oceans, cleansed in a process of distillation, condensed as the atmospheric temperature drops and returned to the earth as rain to make the vegetable growth cycle possible, after which it runs into oceans and lakes and is evaporated again. What drives this system are the forces of gravity and radiant heat from the sun (radiation itself created by gravity) working together in a way that can be called marvellous in the most literal sense. Even more marvellous is the process of photosynthesis by which solar energy is transformed into vegetable mass and vegetable life. Through the catalytic function of chlorophyll electrons are stripped from carbon dioxide and hydrogen dioxide (water) and the basic elements rearranged to create sugars of various kinds (typically $C_6H_{12}O_6$) from which energy is released to enable the plant to push back against the ever-present force of gravity. Where gravity comes from and why it is a permanent force are questions that have exercised physical theorists since Newton (who refused to get involved) but have come within answering distance since Hubble, and the space telescope named after him, have enabled us to wind back cosmic time.

From these very simple examples there is much to learn about nature's energy management that may be applicable to the engineering challenge. The K-gen system is only the start of a different approach to energy theory and practice that may reasonably be expected to become a "new normal" in due course. That lies ahead, of course, but two general remarks may perhaps be worth mentioning as lines for future research. Firstly, if magnetism and gravity, the two permanent forces of nature, can be arranged to act cyclically in opposition, much as heating and cooling activate a Stirling engine, we shall have invented a permanent source of usable energy. Secondly, since an iron bar exerts as much pressure in contracting as it does in expanding, any system which applies and withdraws solar heat to it sequentially is in theory capable of being developed into an engine. These obvious facts will be looked at a little more closely in the following section, for it flows from an important principle.

### 7.4 Three Principles of Energy Management

Three aspects of energy science are of critical importance in framing a K-gen programme. The first involves the principle of trigger energy, that is to say, the use of small amounts of energy to release larger amounts of latent energy.[3] This is principle is seen at work in the heat pump, where low grade heat is augmented by transferring it to a working fluid with low boiling point. The working fluid is then evaporated, compressed and then cooled to release the latent heat of state change, as it passes back to a liquid from gaseous state. This is very elementary science, but making maximum use of this latent energy of phase change is an engineering challenge that has not yet been fully recognized and will be taken up later. In anticipation it may be said here that in ideal conditions the expenditure of 1 kWh of electricity to drive the compressor of the heat pump can release 7 kWh or more in the form of heat – the ratio of 7:1 being its coefficient of efficiency. If that "free" heat can be converted back to electricity we are entering new ground in domestic energy management.

The second scientific principle which bears on energy generation in all its forms is the fact that energy is required to move any system from an equilibrium or ground state and is released when it returns to that state. This may be called the principle of Return to Equilibrium, or RTE. It is foreshadowed in Carnot's insight that a cold sink is as important as a hot source for a heat engine to function and in Tesla's similar but never developed intuition that utilising the differential is at the heart of renewable energy. He illustrated the principle at issue with a thought experiment in which "two metal rods ran from the earth to outer space [creating] a temperature difference between the ends of the rods which … could operate an electric motor."[4] Unfortunately, although the most practical of visionaries, Tesla did not give any details about how the latent power of the temperature differential could actually be translated into electrical power. That the RTE principle is largely unnoticed as a scientific law is due largely to the fact that it overlaps, and in one sense goes against, the second law of thermodynamics. Where the second law is taken to express the seemingly unquestionable fact that hot things spontaneously become cold, the RTE principle would subordinate this to the more general law, that all energetic systems spontaneously seek equilibrium.[5] There are significant issues at stake here, particularly as regard the notion of spontaneity. From the engineer's point of view, however, these metascientific questions may be set aside: the overarching significance is that energy is needed to move any system from an equilibrium state and is released when it returns to that state. This being so, the energy engineer's task can then be seen to be in large part either to identify natural states of disequilibrium or create them artificially in order to tap this latent or potential energy. The importance of this to the renewable energy engineer hardly needs to be stressed.

## References

1. There were also important papers on the theme of engineering and education, the vital role played by engineers in building civilization being a totally blind spot in normal education. In a 2011 survey of 16 to 26 year-olds asked to name a famous engineer, the majority were stumped, and of those who were able to respond, most came up with "Kevin Webster", the fictional car mechanic from the soap opera *Coronation Street*.

2. Gauray Agnihotri, "Germany Struggles with too much Renewable Energy," Zerohedge.com. Retrieved August 18, 2015. The title is somewhat misleading. Wind and solar generated electricity has now made up for some 20% of Germany's needs that had been lost by closing down its nine nuclear plants. Unfortunately, the new supply is largely uncontrolled and often comes when not required by the Grid.

3. The term trigger energy is essentially a transfer to thermodynamics of the category that Arrhenius identified as "activation energy" in chemistry. His insight came from the fact that chemical reactions are speeded,

slowed or halted by ambient temperature, and thus a certain amount of energy is needed to activate the creation or release of molecular bonds. Like gravity, this had been so obvious as to evade notice, and the need for theoretical analysis. Trigger energy is a more general category, and could even be applied to the pressure needed to activate an electrical switch. Its importance becomes evident in designing control systems.

4. "On Electricity", *Electrical Review*, 27/1/1897.

5. Eric Schneider and Dorion Sagan have proposed a somewhat similar principle as a way to bridge the gap between a universal law of entropy and the inescapable fact that life of its very nature is a breaking of the law. In *Into the Cool: Energy Flow, Thermodynamics and Life* (University of Chicago Press, 2005) they propose that spontaneity and ultimately pattern-forming are the consequence of energy moving down a gradient. "The world changes when you view it through the lens of irreversible gradient reduction .... Our reformulated second law presents itself as an active force" (p. 77). RTE is proposed as a more radical reformulation, with greater explanatory value.

# Chapter 8

## The Economic Dimension

*The Feed-in-tariff will change the way householders and communities think about their future energy needs.*

Department of Energy and
Climate Change press release, 1/2/2010

### 8.1 The Bigger Picture

There are really two economic dimensions to microgeneration in general and the K-Gen/E-plus system in particular. The first concerns the impact on the nation's economy that one can foresee if millions of houses are able to draw all or most of the energy they need from renewable resources and link up to feed the National Grid. The second concerns the effects that microgeneration will have on the household budget. It is a fast changing picture, but when everything stabilizes, perhaps twenty years from now, we can anticipate a re-laying of the very foundations of energy economics. If the argument of this book is proved, macrogeneration in its present form, from coal, gas and nuclear, will be on its way out, to be replaced with a mix of microgeneration and mesogeneration of electricity from renewables, which will fulfil all domestic and industrial needs. On the way a high degree of economic turbulence is inevitable. It is widely agreed by economists that a tipping point will be reached as the world's oil supply drops below a critical figure and the price of oil reaches $150-160 a barrel, assuming the 2015 value of the dollar. At this price a rapid squeeze on all aspects of economic production will begin and a global recession of unparalleled depth and intensity will set it. With this as background, it can be predicted that if an ordinary house can be designed that will produce, say, 50% more than its own energy requirements, a revolution in energy production and use will be jump-started. What works for a house can work for a factory, and even more efficiently. So we can expect to see dedicated installations, small "energy farms" harvesting solar and wind energy on the medium scale. Microgeneration and mesogeneration will come together and the whole energy landscape will be changed.

There is every reason to anticipate a new kind of energy industry starting up, because the government's current Feed-in Tariff (FIT) is available not only for domestic residences but for commercial suppliers, albeit at a lower buy-in figure The movement in this direction is already under way, with mesogeneration quite suddenly taking on a new importance. At February 2012 there were 85 planning permissions in the pipeline for 5 megawatt (5,000 kW) or greater solar farms. However, the government seems to have been taken by surprise by its success in stimulating commercial interest, for this number alone will use up virtually all the cash it initially set aside to promote solar energy, and the Treasury Department is unlikely to look favourably on boosting it, when it is attempting to cut back on everything else. Domestic subsidies have already been halved and one can only hope that they will not be cut to a level where they no longer act as incentives, and the original purpose of offering them is forgotten.

If an intelligent strategy should eventually be activated, for both meso- and micro-generation, we may expect to see major changes in the National Grid, and a U-turn in present planning policy, which is biasing design of a smart Grid to maximize utilisation by giant wind farms, as earlier indicated. With micro- and mesogeneration playing a substantial and growing part, what could happen is that the Grid, which is now essentially a trunk and branch structure of centralized distribution from a few very large installations, would become a national network operating on three interlinked levels. We can envisage, for instance, perhaps twenty regional networks, each with perhaps a million small and medium scale generating installations and a few large scale installations all forming a single flexible system. This type of scenario is without doubt some way in the future and macrogeneration may by then be almost entirely in the form of a small number of thorium-fuelled reactors, unless the government persists with its present makeshift policy of totally subsidised uranium-fuelled production plus fracked gas. The situation is extremely fluid, and current developments at all levels make more detailed prediction impossible. One very interesting possibility arises from price competition, for to the extent that micro- or meso-generated electricity may prove cheaper than large power stations of any kind, because they have much smaller capital costs, the government may have to step in to subsidize all macrogeneration, not just uranium-fuelled, as at present, if only to ensure a smoothing of supply and contingency back-up. This line of thinking leads continually to the need for a centralized authority, with a genuine think tank at its core, which will address the requirements of the country rather than the profitability of a few giant energy companies. Whatever the future holds, there will be a great deal of adjustment needed as a national supply system settles into a steady state powered almost entirely by renewable energy sources.

Although macrogeneration from renewables has had to be excluded from the present work, it is worth noting here that huge solar farms are already functioning in Australia, Spain and America and the development of this sector is going ahead rapidly. It is very difficult to establish whether or not any of them will have long-term profitability, as various kinds of subsidy and ongoing expense for back-up electricity supply are buried in their accounts. Since the early 1980's the Mojave Desert in California has become the home of several kinds of solar generating installations, ranging from acres of photovoltaic panels to solar troughs and dishes, and is in effect a testing ground for different large scale technologies. The Blythe plant uses solar troughs to superheat oil which is then used to drive generators through steam turbines. Some designs use parabolic mirrors that focus the sun's rays on the heat face of Stirling engines, with back-up supply at night from fossil-fuelled installations. Other designs use molten salt technology, which is becoming increasingly popular, because the stored heat can be used when the sun goes down. In Spain the Gemasolar plant using this technology was built in 2010 at a cost of £260m to supply 25,000 homes. Solar energy is reflected from some 2,600 flat mirrors, spread over an area the size of twenty football pitches, which are angled to focus onto a tower containing salt. The salt liquefies at 850°C and as it cools, releases the heat of liquefaction, enabling steam power to be generated both day and night, thus obviating the need for back-up supply from conventional sources. The Gemasolar website claims that the system has 16 hours of reserve power through this storage method, but it is not clear at this point whether this figure is a summer maximum, when some heat must be dumped, or if it applies to the short daylight hours of winter, when sunlight is not so strong. At first glance it seems that there is nowhere near enough storage for winter demands and, if this is so, there will need to be a conventional coal or gas-fired back up facility, the cost of which has, presumably, not been factored in.

The financial data must have a wide margin of uncertainty until the plant has been operating for several years, but as an initial point of reference, it can be seen that the estimated gross capital cost of the Gemasolar installation would be about £1,000 per home. If this approximate figure is borne out over time, it would be very economical indeed but, of course, the solar environment in Manchester is very different from that of the Mojave Desert or Andalusia. Solar energy schemes on this scale may be cost-effective in the south and south east of the UK, which is relatively well endowed with sunshine, but as one goes north and west, their cost-effectiveness will inevitably diminish to zero.

The Desertec Industrial Initiative (DII), already mentioned is on a scale which dwarfs all the above and, if implemented, will alter the energy landscape of Europe. The big question is, can it be implemented, and this brings with it all manner of unsuspected questions, too many to be even sketched out here. On the one hand, the fact that seed capital of £400 million has been pledged by the World Bank and that a dozen of the world largest transnational corporations have come together to design and drive the project would suggest that the risk/return ratio has been carefully considered. Among these companies are Siemens and the civil engineering giant ABB, but also the giant reinsurance company Munich Re, surely the global expert in financial risk assessment. The grand plan is to build solar thermal plants across North Africa to the Arabian peninsula connected to a high voltage transmission system which will take most of the electrical output across the Mediterranean. On the other hand, the planned locations are arid as well as sunny and it does not appear from the information provided in the DII *Whitebook* that the problem raised by the need for water has been adequately resolved, and talk of using the abundance of energy for desalination plants seems at this point to be little more than handwaving. It is worth noting that Hermann Scheer was very much opposed to the project, and no one had a better grasp of global and social needs than he.[1] A sixteen page, but still rather thin, overview of the viability of the project can be found on the Internet.[2] It is tentatively supportive, but does not enter into serious questions of international finance and geopolitics that are inseparable in a project of this size and nature. One gets the feeling that as publicity for this venture has gone very quiet for a year or two that unsuspected problems have arisen to cast doubt on the concept itself.

Whether or not this speculation is correct, the giant transnational engineering companies involved have the ear of government, and inevitably bias national planning towards large and very large scale technological solutions. Given that they work in synergy with the large international banks as well, there is a continuing danger that small scale solutions to the world's energy problems will simply not be seen, or will be dismissed as the unreal hopes of eco-dreamers. In the end the market will decide, unless governments intervene to favour big business. If the unit cost to the Grid for microgenerated power is lower than for macrogenerated, we can expect to see an energy revolution.

Other kinds of macro-installations harnessing renewable energy, such as tidal barrages, are unlikely to play a significant part in the larger economic picture, if only because there are few suitable sites for development. Commercial development of submerged turbines in rivers and estuaries is as yet an unknown quantity, but test installations seem to encounter the problem of weak and undependable flow pressure. However, while there have been numerous false starts in the past and wildly over-enthusiastic forecasts, the fact that the government is now providing financial backing for a "marine energy park in a bid to speed up the commercial expansion of the wave and tidal industry from 2020" suggests that past research and testing may be starting to pay off. This watery "park" is intended to stretch from Bristol to Cornwall and the Scilly Isles. The Reuters Report of 23/1/2012 from which the quotation is taken also contains an assessment of its economic value by Greg Barker, the then Minister for Energy and Climate, "Marine power has huge potential in the UK, not just in contributing to a greener electricity supply and cutting emissions but in supporting thousands of jobs in a sector worth a possible 15 billion pounds to the economy in 2050." One must wonder on what basis he has done calculations that extend so far into a future when the global and national energy situation even in 2030 is so completely unknown. However, these uncertainties work both ways, and it would be premature to dismiss such a scheme as unrealistic, despite the poor track record of such ventures on the small scale.

As regards the continued contribution of giant wind farms, while political forces and "sweetheart" subsidy deals may keep the industry going for some years to come, in the long run, there is no reason to think they will be able to compete on price. As just remarked, all will hinge on the unit cost of baseload electricity, available permanently and on demand, and if microgeneration and mesogeneration can meet this demand cost-effectively, they will become a future standard. This simple scenario becomes more complex, however, when the enormous capital cost of large scale renewable energy programmes is taken into consideration in a world of finance that becomes more chaotic all the time and where almost all banks are cutting back on their loan books in order to recapitalize. All these factors add up to create unwillingness on the part of either government or the private sector to finance a comprehensive strategy, even though that clearly is what the nation, and the planet, require. Dramatic cutbacks in government expenditure across the board, brought about by the looming avalanche of interest to be paid on the national debt, mean that large scale investment in energy is unlikely, despite the ultimate payback. This in turn suggests that the emphasis will eventually be put on domestic scale energy projects, for the simple reason that capital investment can then come mostly from the householders' savings and income. The prospect of a largely self-funded microgeneration initiative must look very attractive to governments suffering from severe financial constraints. This political aspect of microgeneration economics can hardly be over-emphasized, for government policy must eventually be decided by millions of citizens voting with their cheque books.

**8.2   The Economics of Domestic Energy**

To get a grip on the slippery economics of domestic electricity generation in the UK it is necessary to understand that the government strategic incentive plan is actually three separate plans operated by apparently non-communicating bodies. As the quotation at the head of this chapter illustrates, the UK government, through the newly renamed Department of Energy and Climate Change, clearly had the intention of initiating an energy revolution when it introduced its direct financial support for microgeneration through photovoltaic panels in February, 2010. Enthusiasm was running high, and it was promoted as a "blueprint published for the world's first incentive scheme for renewable heat". That was not strictly true, in fact, since Germany already had a well-developed feed-in policy and ranks of solar panels can be seen along motorways especially in southern Germany, where farmers have taken full advantage of the financial benefits, which amounted to a quite astonishing $130 billion.[3] Significantly, these are now being phased out. The FIT scheme in Britain offered no capital subsidy but very generous and guaranteed prices for electricity produced by a household for its own use as well as for a surplus fed into the Grid, regardless of whether or not the Grid could use it when supplied. Subsidy for production of hot water through solar thermal panels and other devices was offered through the Renewable Heat Initiative (RHI), which was developed by a newly constituted body, the Renewable Energy Association and administered by Ofgem. House insulation was at the same time promoted by several bodies, including most local councils and a wide spectrum of commercial enterprises (including the retail grocer Tesco) and was funded by the energy utilities. This latter fact is a curious anomaly, since the more successful a campaign for house insulation turned out to be, the more their profits would suffer, but this is just one aspect of a scene which is lacking in co-ordination and which calls for a unified command structure, as the earlier quotation from Jeremy Leggett has made clear. The Energy Saving Trust was specially set up as an online information hub, and no doubt fulfils this role in a narrow sense, but the urgency and complexity of the situation really calls for a single body with executive and planning functions. While there is sporadic recognition of the need for focused, properly funded and coordinated action to deal with the energy crisis into which we are now entering, it seems that other priorities always seem to prevail. The Energy Manifesto of the Conservative party for the 2010 election promised a specified grant of £6,500 per household for renewable energy installations, which would have brought a degree of simplification to the present confused scene, if implemented, but it went the way of most election promises. Does anyone now remember it?

From the perspective of the householder, the economic costs and gains are difficult to assess, and have become virtually impossible since the FIT subsidies were abruptly halved, and then cut again, with no assurance that this minimal incentive figure would be the bottom. The Renewable Heat Incentive, which covered both home-owners and businesses, was provisional from the start, with firmer arrangements coming into operation in 2013. Grants are variable and certain classes (e.g., pensioners) can claim the full cost of both loft and cavity insulation. Since few would have to pay more than £200 towards the capital cost, and since both FIT and RIH subsidies are dependent on having a fully insulated house, there is a considerable incentive to take this first step towards economizing on fuel but, even so, fewer than half the houses in the country have taken advantage of it. The cost-benefit to the householder of adopting all three schemes may be calculated roughly as follows, but, as will be seen, it is very much a moving target.

The house insulation grant is the easiest to compute, for if the householder's contribution, probably a maximum of £400, is set against a notional saving on heating bills of at least £150 per year, it can be seen that the payback time for investment will be as little as three years. In very rough terms the theoretical return on investment could hardly be less than 30%, though anecdotal reports suggest that the real saving is far less. One suspects that the payback figure was deduced from tables of insulating materials, whereas the ordinary house loses heat from badly fitting doors and windows and floors which have no insulation at all other than a carpet.

The feed-in tariff was clearly designed initially as a powerful incentive, indeed as an offer that could hardly be refused, since home-produced electricity would be purchased by the Grid at 41.3p per kilowatthour (for retrofitted installations) as against the average cost to the consumer of about 12-14p; this figure was index-linked to inflation for 20-25 years and the cash income, paid quarterly, would be tax-free. Against a capital cost at that time of about £14,000 for a 4 kilowatt PV panel installation, the nominal return would be about 5-6%. However, given that it would be tax-free and inflation-proofed, the real return could well have been in excess of 15%. In December 2011 the Department of Energy slashed the buy-in return from 41.3p per unit to 21p, which, on the face of things, would halve the long-term return to 7%, still a very acceptable figure in itself and even more valuable if inflation were to get out of hand. However, the actual loss to the householder of the reduced subsidy was much less than appears on the surface, since the average cost of a 4kW installation has come down dramatically, from about £14,000 to £9,000. Indeed, this reduction in capital cost was a major consideration in the government's decision to stop an incentive from becoming a bonanza. Simple arithmetic enables the real return to be calculated at about 10%, still a very worthwhile figure.

The FIT initiative covers wind-generated electricity as well as PV panels, and its broad structure may be shown in the following abbreviated table, showing both the original and current figures. (The tariff is normally guaranteed for 20 years.) The "standalone" item is for PV panels that can be located in the garden or elsewhere in any number, thus offering the householder the possibility of becoming a small scale commercial electricity supplier. In passing, it is perhaps worth commenting on the way in which decimal points of a penny are used, even though so small a distinction is meaningless in practice. One must ask what kind of mindset was behind the whole scheme, for anyone with the slightest experience of marketing would have gone for big round and memorable figures. So too one can see a lack of focus in the differential between small and large scale generation, both PV and wind. If it is electricity that is wanted, the size of the installation is irrelevant. Indeed, if there were to be a higher buy-in price for larger installations, it would encourage the householder to make a bigger investment.

Photovoltaic Panels

| Rated Capacity | Buy-in Price per kWh | |
|---|---|---|
| 4kW | initial | current |
| | 41.3p | 14.9p |
| Above 10kWh the tariff reduces to 12.57p | | |
| Standalone (any size) | initial | current |
| | 29.3p | 6.85p |

The government's wavering commitment to domestic scale renewable energy seems to have disappeared altogether, with a "consultation paper" of August 27, 2015 which proposes to cut the solar voltaic feed-in tariff by 87% to 1.63p, effectively a token payment.

## 8.21 Solar Thermal Panels and Heat Pumps
As against the FIT scheme, which encourages the production of *electricity* with ongoing payments but offers no capital grants, the quite separate Renewable Heat Initiative (RHI) is directed to the production of *heat* (effectively hot water) and gives substantial grants for equipment, which covers solar thermal panels, air and ground source heat pumps and biomass boilers. The figures at April, 2015 (from *Which?* Magazine) for a 3-bed detached house are as follows:

Solar thermal panels - £383 per unit
Air source heat pump - £1,679
Ground source heat pump - £4,324

Across the whole range of houses the RHI website estimates a return on investment of "between 7% and 8%" from savings in fuel bills. The situation is, however, more complex than this for several reasons. Included in the RHI are wood-fuelled heating installations (£950) and ongoing subsidies of a "fixed amount per year ... to anyone who is using a renewable alternative", based on either "the exact amount of heat produced - which means you'll need a meter fitted to your system - or on the amount of heat the installation is anticipated to produce." In 2013 the method of calculation was changed, the subsidy for solar thermal units being set as "at least 19.2p per kWh of solar thermal energy" but based on the number of people in the house, so that a single occupant would typically receive £1,150 over seven years and a house with six occupants almost three times this amount. The logic behind this appears to be that since it is impossible to measure the heat that is produced or needed, a "deemed performance" can be calculated as a proportion to the number of people needing hot water or the comfort of central heating. This method of reckoning would stop individuals from just adding solar thermal panels to their house simply to get the subsidy. The other aspect of government thinking that appears in the change is that the subsidy for solar thermal panels and heat pumps is now phased over seven years.

This line of enquiry leads to the question: who pays? At present there is a limited pot of government money allocated to both the FIT and RHI schemes, but whereas the ongoing cost of the former is met from a hidden levy in everyone's electricity bill, there is no possibility of rewarding the householder in the same way for producing hot water, and so future rewards must come from general taxation. There has been much guesswork about the actual figures of the hidden levy on electricity bills, but little objection to the fact that those households which have no energy saving installation are paying more for their electricity in order to subsidize those who have. Estimates of "how much more" have ranged from £12 per year to the definitive £80 given by the Energy Minister. This rather alarming figure may have been deliberately produced out of a hat as a hidden incentive to persuade householders to get on the subsidy bandwagon, rather than pay for those who were already on it. No firm figure is possible, for the obvious reason that as the government's incentives create more home-produced electricity, the burden of paying for them will fall on a reducing number of those householders who have not taken up the offer. The fact that the RHI offers a subsidy for wood burning stoves indicates that the creators of the scheme were primarily concerned with increasing the national electricity supply rather than with reducing carbon emissions.

In the tangle of incentives and disincentives, with different energy supply companies offering a variety of rates and their own incentives, Ofgem has suggested that it may be necessary to have a single, state-controlled energy buyer, but the logic that drives this hinted proposal leads to the conclusion that it will be necessary, if only to standardize the cost and rewards to householders nationwide. As this comes home, unseen social consequences of the energy crisis start to unfold, for it is hard to see how this could be done without taking back energy production and distribution into public ownership, in practice if not technically. After a flood tide of privatisation across the globe for thirty years, it is quite possible that this may herald a new economic revolution.

The cost of an efficient K-gen/E-plus system cannot be assessed until several prototype houses have been built. When integrated in the design of a new-build house, the additional cost can be absorbed in a mortgage and in this case there is every reason to think that monthly instalments would be more than offset by savings on energy bills and profit from selling surplus electricity. A dedicated system could be expected to add between £20,000 and £30,000 to the cost of the house (the figures at this point can hardly be more precise) but will add significantly to its value, and despite the daunting size of the initial investment, it would seem to be good value for money if paid for over the lifetime of a mortgage.

When it comes to a retrofitted system, however, this method of valuation is not so easy to apply, since every house will have its own parameters and in the worst case (say, a heavily shaded apartment block) may not generate a surplus of electricity. A retrofitted property may or may not be remortgageable to spread the capital cost, and the government has recognized this anomaly by setting up a scheme within the Energy Saving Trust that treats a retrofitted system as a mortgageable investment in a so-named PAYS (Pay as You Save) plan. At present PAYS is being piloted through partnership between the government and five local councils and commercial enterprises, among them Birmingham City Council, British Gas and B & Q. As at January 2013, it is providing £4 million of initial funding, with a view to testing the viability of the arrangement – essentially finding out if the visible monthly saving on electricity and gas offsets the long term mortgage payments. Clearly, the hope is that councils and commercial enterprises, and perhaps building societies later, will welcome the opportunity to go into the business of long-term mortgage-lending on retrofitted microgeneration installations, as they become aware of a secure profit margin. Given all the present uncertainties, testing the water through a small scale pilot scheme is an intelligent first step.

## References

1. Hermann Scheer, "European power from the desert is a *Fata Morgana*." An article on his 2011 website.

2. Sarah Irving, "The Desertec Mirage: The Validity of DII Skepticism," *Atlantic International Studies Organization*, May 2012.

3. The figure is from a Report published by Ruhr University, February, 2012. Germany's experience illustrates the pitfalls of a full-on renewables strategy driven by financial incentives. The two most obvious are a recurrent theme in the present document. Firstly, someone has to pay for subsidies and that "someone" is not a government with unlimited resources, but the taxpayer. The average German consumer now faces an increase of $260 in their annual energy bill (*Money Week*, 24/2/12). Secondly, because of the intermittent nature of solar and wind energy, there must always be conventional back-up generation plant of the same capacity available for extreme circumstances, a point already made in relation to wind farms. The great engineering challenge, and the "elephant in the room" in every government plan for renewables is energy storage, short and long term.

# Part Three

# The K-gen System

*One day man will connect his
apparatus to the wheelwork of
the universe, and the very forces
that motivate the planets in their
orbits and cause them to rotate
will rotate his own machinery*

Nikolai Tesla

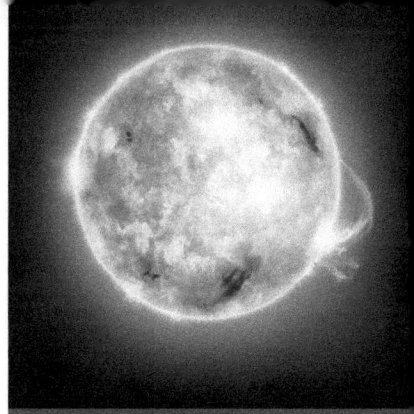

# Chapter 9

## The K-gen System: Parts and Functions

*When we learn how to store electricity, we will cease being apes; until then we are tailless orangutans ... There must surely come a time when heat and power will be stored in unlimited quantities in every community, all gathered by natural forces.*

Thomas Edison [1]

### 9.1 Going to Proof of Concept

This part of the book develops the themes of the K-gen and E-plus systems whose function and rationale have been dealt with in a general way in Chapter 6. It will zoom in, as it were, on the five-fold strategy by which the K-gen is defined, culminating in a schematic of the system and the outline of a research programme leading to proof of concept. It might, in fact, be more appropriate to use the plural "research programmes" since, as will be seen, several of the component parts will call for design and testing before the system as a whole can be assembled and tested. Chapter 10 will deal more specifically with the architectural issues entailed in an E-plus project. The starting point for any of the separate research programmes must be a thorough survey of the state of the art in each of five fields, and a search of the existing literature will itself be a substantial task.

### 9.2 Energy Collection

There are two obvious sources of renewable energy, namely solar and wind power, plus the not so obvious heat pump.

### 9.21 Solar Energy

Photovoltaic Panels

Recapping information in Chapter 6, there are two well developed systems in use to collect solar energy, namely photovoltaic (PV) and solar thermal panels, The most obvious advantage of PV panels is that they convert sunlight directly to electrical current, which can be fed (after inversion to AC) straight into the Grid. Solar thermal panels, by contrast, are used to heat water, thus reducing demand on gas or electricity supply used for this purpose. PV panels are based on cells of silicon, which both collect and convert the sunlight, much as chlorophyll molecules operate in plants. There are several competing types of cell, including standard, thin film and spray on, the latter two being amenable to mass production. A recently introduced type of photovoltaic "mini" cell claims enhanced performance by using collecting lenses. There is also a very simple system that uses angled mirrors to increase the amount of sunlight falling on the panel. As there is active research and development on all these fronts, selection of the most cost-efficient at this point is not possible, and a comprehensive assessment of the benefits and drawbacks would call for substantial research. It is, however, worth mentioning that the efficiency of PV panels falls off as they heat up, so the best design must have provision for cooling. Bearing in mind all these variables, the solar energy collected by a one square metre panel could range from 800 watts at summer peak to 120 watts on a bright December day - of which there are not many in the UK. The actual electricity generated is, however, only a fraction of this, typically about twelve per cent, though constantly increasing with research, thus providing at best, and irregularly, about 80 watts in the summer and a negligible amount in the winter months, when most needed.

## Solar Thermal Panels

Solar thermal panels producing only hot water, are inherently more efficient, typically converting 80-90% of the solar energy they receive into heat in the form of hot water. Evacuated tube collectors, invented in the 1980's in China, offer much greater efficiencies than the standard flat plate design. They work well all the year round and even when the sun is not shining generate substantial heat from ambient light. Surprising as it may seem, winter sun on a very cold day is sufficient to heat water to 40-50 degrees, which is adequate for a low temperature heating system, on which more will be said below. The obvious drawback of evacuated tube collectors is their initial high cost, nearly twice that of flat plate panels, but they are

so superior in critical respects that they would be the first choice for
an E-plus house.[2]

## 9.22 Wind Energy

Wind is a notoriously unpredictable source of energy, almost never
cost-effective on a domestic scale in an urban environment and often creating noise pollution to boot. The most comprehensive data
on urban domestic wind-powered electricity generation in the UK
is to be found in the Warwick Wind Trials, and the results are not
encouraging. From records of thirty installations, both roof mounted and free standing, compiled between October 2007 and 2008,
the gross average amount of electrical power was no more than 78
kWh per year. This represents a saving in electricity of about £10
per annum on a capital cost of anything up to £2,500. Several of
the wind turbines had to be switched off at night because their noise
disturbed sleep, and some were switched off altogether because of
noise. There was an unexpected amount of down time through mechanical failure or faults calling for maintenance. It may reasonably
be concluded that with more reliable machines, the electrical output
could probably have been doubled, but that still would not make
the capital investment worthwhile. There were wide variations in
performance, reflecting partly the efficiencies of the different models and partly the complex wind patterns and flat spots in a built
up area. The few which were free standing, higher than roof level
and in large gardens or other clear space, performed significantly
better than those which were not. All models appear to be conventional horizontally mounted, un-ducted, propeller-type turbines
and hence the final figures cannot be taken as conclusive, since there
are several alternative types already on the market, plus experimental
models based on new principles which will call for attention later in
this chapter. Wind turbines apply the same principles as were first
used in ancient Persia in 200 BC. The issue of wind turbine design,
its history and its failures offers more than enough material for a
university degree course and the salient facts offered here must be
very selective and restricted to harvesting wind energy on a domestic
scale. That said, it will become apparent that while there are oppor-

tunities at the level of microgeneration, mesogeneration is likely to
offer even more.

As regards the basic science, it is well realised today, as it was not
in the past, that the energy in moving air has two components, not
easily separable, namely kinetic and inherent pressure. The ability of
a sailing ship to tack into the wind depends upon the latter, whereas
commonsense would suggest that the ship would simply be blown
backward. The collectable energy of wind increases as the cube of
its speed, an important fact that should encourage wind turbine designs that can make use of high wind, as against standard models,
which must feather the blades in high winds to avoid damage. Five
other principles call for mention, the first being the obvious, but
too easily ignored, fact that a decrease in the back pressure of a wind
turbine is the equivalent of an increase in the pressure of the impacting wind. Secondly, there has been major new understanding of the
properties of air in motion, most notably the so-called Coanda effect
which comes into operation when moving air entrains into motion
static air at its boundaries. (The "magic" of the Dyson "Air Multi-
plier" cooling fan relies on this.) Thirdly, the force of air in motion
regenerates after it meets an obstacle and flows round it. Fourthly,
and related, is the spillover effect, which means that a flat surface
facing into the wind cannot collect all its theoretical energy, since
much of it flows over the edges. Fifthly, air in circular motion creates
a tornado effect which automatically concentrates wind energy at its
base. All these principle need to be kept in mind when considering
how wind energy can be harnessed most efficiently.

The efficiency of a wind turbine is generally taken to be given by
the Betz law, which limits it to 59.3% of the energy of the impacting wind. This, however, assumes two factors which apply only to
simple windmill design, namely that the collecting area is defined
by, and limited to, the area swept by the blades and that there is
a constant back pressure. The latter puts a limit on the amount of
air that can be, as it were, pushed through. Conventional design
makes no attempt to decrease the back pressure behind a turbine
in order to augment the energy of the incoming with suction. The
Betz limit, taken with the fact that windspeed increases dramatically

with height above ground level, has had a dominant effect on wind turbine design, leading to ever taller installations (often more than 100 metres) and ever longer (and heavier) blades. The inefficiencies and material cost of this approach are rarely brought under scrutiny.

The most significant division in conventional wind turbines is taken to be between horizontally and vertically axis mounting (HAWT and VAWT). These are then subdivided in different ways. Domestic HAWT installations are usually open rotor, but cowled or ducted models are growing in popularity, especially at mesogeneration scale. These are sometimes referred to as DAWT - diffuser augmented wind turbines - and the principle behind them can be seen in this simple diagram.

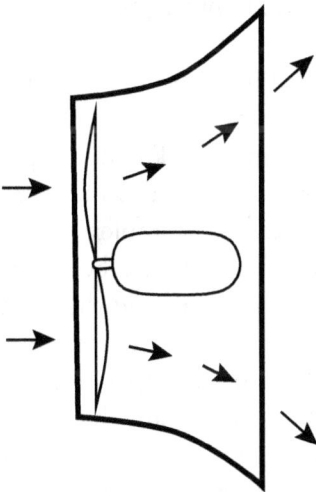

Fig. 4   Diffuser augmented cowled wind turbine (DAWT)

DAWT systems offer an advantage in that the wind which passes between the cowl and blade tips diffuses as it exits through the wider rear aperture, thus decreasing back pressure to give a substantial overall gain. The "Wind Tamer" design, with a maximum diameter of about two metres and a rated output of 15 kW, claims a rather surprising output of 240 watts per square foot of rotor sweep in a thirty mile per hour wind, as against 40 watts from an open rotor. A thirty mph wind, it may be noted, although labelled only "a fresh breeze" on the Beaufort Scale, is not a frequent occurrence at ground level in non-exposed areas, and for practical purposes is judged to be when umbrellas become unusable. A proper specification should include the height above ground as well as the notional windspeed. Another advantage comes from the more robust construction of the DAWT system and the smaller diameters of its blades, which enables it to operate at windspeeds when open rotor models must be switched off to avoid structural damage. All HAWT's, whether open or cowled, have a potential disadvantage in that they must be pointed into the wind for maximum efficiency and produce no electricity at all when they are side-on. This is easily overcome in small installations by adding a wind vane, which can be as simple as a weather cock, but large wind turbines require an auxiliary motor to perform this function, which not only uses up electricity but loses efficiency through time lag.

By contrast, vertically mounted wind turbines have the great advantage of collecting the power of the wind from whatever direction it might blow. Blade design is all important in VAWT's and ranges from simple paddles that provide angular momentum as the wind strikes them through scoops (Savonius model) to complex helical designs using airfoil principles (Darrieus). VAWT's have been undergoing development for nearly a century and a half and there are many working models as well as a wealth of technical data available. The pros and cons as regards macro- and meso-scale operation are too complex to summarise, and it is perhaps significant that there is negligible promotion of micro-scale models for domestic generation. There are, however, several models which are marketed for use on small boats to trickle charge heavy duty marine batteries. The Renewable Energy UK (REUK) website offers for about £150 a 50 watt rated VAWT with telescopic mounting pole, with dimensions about a metre high and diameter of 200mm. Even though its real output is likely to be no more than 20 or 30 watts, this opens up possibilities for domestic microgeneration, since for less than £1,000, half a dozen roof or chimney mounted installations feeding into a high efficiency battery opens up new vistas for electrical storage, which will be looked at more closely in Section 9.52 below. An intermittent input of 80 watts on a windy day could provide half a kilowatthour of energy, and if this can be stored until required, it would be very useful and cheap.

Two promising developments in domestic wind turbines appear to lie in applying the principles governing the behaviour of air in circular motion (the tornado wind effect) and in horizontal roof top turbines which use the roof as a wind collecting device. Information on the former can be found through an Internet search for TWECS (Tornado Wind Effect Conversion System), which seems to have become a generic name. Horizontally mounted roof turbines can simply be bolted onto to the peak of an existing house or approached as an architectural challenge, and will be looked at below. TWECS development has been going on for some thirty years, and has been held back partly for lack of funding and partly because successive working models turned out to have critical flaws. A paper entitled "A New Approach to the Tornado Wind Energy Conversion System" presented at the 7th World Wind Energy Conference, 2008 offers a good historical overview and promise for the future. The basic principle, as shown in the diagram below, is to create a vortex, and thus a pressure differential, by forcing incoming air into circular motion.

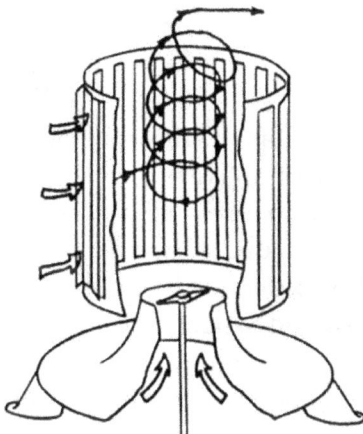

Fig. 5   Tornado wind turbine system (TWECS)

The principle behind the concept is that air enters through vertical louvres in a cylindrical construction, and is forced upward and into circular motion before it exits at the top, thus creating a low pressure area at the top and a high pressure area at the bottom, where the system is connected to a turbine. The difficulties of going from design to proof of concept have proved considerable, and do not yet seem to have been overcome. One of them has been that although a TWECS does not need to face into the wind, since the input louvres go all round the cylinder, it must have some kind of automatic device for closing the louvres on the lee side. The inventor of the system, Dr James Yen estimated in 1976 that a working model would quadruple the energy output of any existing DAWT system, to 1000 watts per square foot of rotor sweep. It is perfectly possible that this claim is too modest, since the initial designs were conceived before magnetic levitation bearings had entered into mainstream engineering. When a maglev bearing is incorporated, frictional loss from conventional bearings can be effectively eliminated. After various vicissitudes the system has been taken up by the Buckminster Fuller Institute and their endorsement promises to bring wider publicity and financial assistance for development.

Roof-peak mounted HAWTs appear to be one of those inventions which suddenly appear in several places at once. Three examples, which may not be exhaustive, are Windpods in Australia, Ridgeblade, a Yorkshire based company and California Wind Systems. All three models are cross flow, helical bladed and with conventional bearings. Significantly, Windpods markets its HAWT on its own merits as a freestanding installation, offering roof mounting only as an option, while the other two emphasize the increased efficiency gained by using the roof of the house to increase the collecting area. Mounting the turbine on the peak of the house not only offers this critical advantage but also accelerates the wind as it flows upward into the turbine. California Wind Systems claim an increase in efficiency (over the freestanding turbine) of 7.5 times. While this figure is plausible and well worth having, it will surely be an optimum, obtainable only when the roof of a house happens to be facing into the prevailing wind. The Ridgeblade website claims that it has been wind tunnel tested up to 100 mph and that the smallest model (6.5m) will produce a substantial annual output of 5,800 kilowatthours. This may be compared with the very low figures given above from the Warwick Wind Trial. California Wind Systems offer a refreshing "grid tied plug and play installation" for $5,000 including inverter, for a neat box measuring 90cm by 2.8m. The Windpods model (not available for domestic use at the time of enquiry) has an unspecified cost, other than A$3,000 installation and grid connect, but offers A$4,500 per kw installed. It is not immediately clear how economical this would be.

What is missing in the promotional material of all three companies is the potential for erecting the system not on the roof of a house but on a dedicated structure. Granted that this would have to be at roof height to gain the same benefit of wind speed, such an approach would offer several clear advantages. Firstly, the structure could be designed to maximize collection, using side walls to funnel the wind, a feature that will be proposed later as a novel element in the E-plus architecture. Secondly, the "roof element" could be pitched at an optimum angle for amplifying the wind speed. Thirdly, it could be extended downwards to increase the collecting area. (This could be done deliberately in house design by creating an artificial roof extension, effectively a very deep eaves and soffits.) Thirdly, the turbine could in principle be mounted on a turntable in order to face it into the wind, thus maximizing its efficiency. This would be a serious engineering challenge, since a servo motor and mechanism would be required, possibly subtracting a significant amount of energy from the whole and adding to the expense.

### 9.23 The Heat Pump Principle
The heat pump is an essential in the K-gen and E-plus systems. A normal heat pump can be expected to have a coefficient of performance (COP) of three to five, that is to say, it amplifies every unit of electrical energy used for input into three or five energy equivalent units of heat. This figure can be significantly increased with a little engineering imagination. The apparent creation of energy does not break the first law of thermodynamics (that energy can be neither

created nor destroyed) since part of the input is in reality "activation" or "trigger" energy, a small amount of which serves to release the greater amount of latent energy contained within the working fluid or refrigerant. The working fluid, primarily chosen for its low boiling point, is the substance which absorbs, holds and releases the latent energy of state change as it is heated from liquid to gas and then cooled back again to liquid. This is achieved by compressing the working fluid in vapour state, then allowing it to expand. As with the steam engine, the power of expansion of the working fluid as it changes to vapour can be used to drive machinery or generate electricity. Electricity is consumed by the pump which does the compression, and the heat energy released as the compressed vapour cools back into liquid form is "captured" by means of a heat exchanger. The "waste" from this process is cold air, but the system can be reversed, as in the domestic refrigerator, so that it produces cold air as a product and warm air as waste. The scope for development as part of both a micro and meso system is considerable and efficiency can in principle be improved by using the energy of sunlight to drive the process. The following diagram illustrates very simply the science behind the engineering. While it may seem logical to assume that adding heat steadily to water will push it up to 100°C, when it suddenly becomes steam (strictly, water vapour), the reality is quite different. Change of state of this kind occurs as the energetic bonds keeping the molecules close together are broken in an on-off or all-or-nothing moment, and the bonds can resist snapping for some time. The result is that one can heat water and watch the temperature rise from, say, 90° to 99°, where it will stick for several seconds, as the heat being added does not raise the temperature but weakens the molecular bonds until, suddenly, they snap and water molecules disperse as steam. When the steam condenses back to a liquid state, the excess heat is absorbed by the surrounding air.

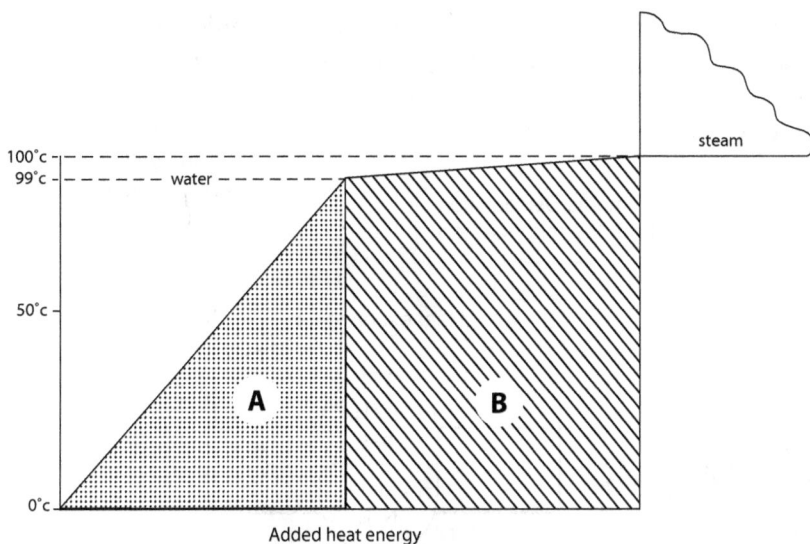

Fig. 6  State change and latent energy
(illustrative only, no scale of any kind)

**A** represents the amount of heat energy required to take water from zero to 99 degrees.

**B** represents the amount of heat energy required to change water at 99 degrees from liquid to gaseous state

The hatched area **B** represents the latent heat of vaporisation. This energy is available for use when the steam cools back to liquid state. How to use it is the engineering challenge.

The Australian government and National University have invested significantly in research into this process, with the intention of converting abundant sunshine into baseload electricity. A 230 million dollar project is underway in Whyalla, South Australia, using solar collecting mirrors to gasify liquid ammonia and re-liquefy it in a closed loop before starting the expansion/evaporation cycle again.[3] The following diagram shows the broad principle. The molecules of liquid ammonia are dissociated by the heat of the sun, then compressed and allowed to expand, in the process driving a turbine which in turn drives an electric generator. As the ammonia in gaseous state expands it condenses back to liquid and is pumped back to the reactor, where the process starts again in a continuous cycle.

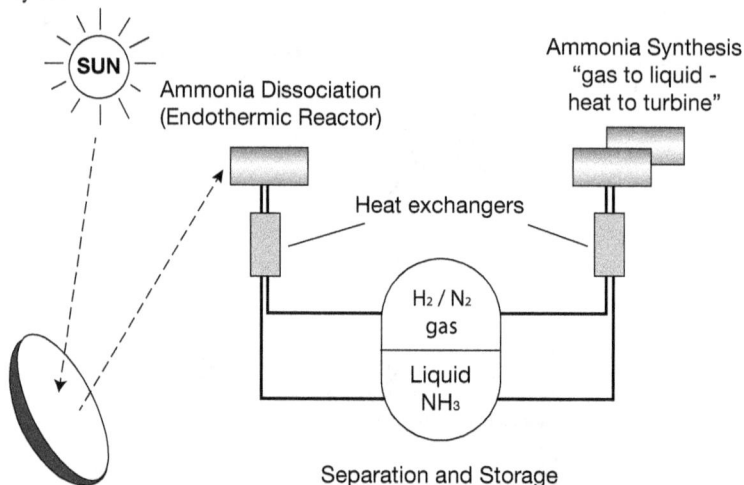

Fig. 7   Closed loop state-change energy generation

How far the Australian experience can be replicated in less sunny climes is a question that needs to be asked.

Heat pump efficiency is the inverse of the difference between input and output temperature and can thus be increased by bringing them closer. Thus if the input temperature can be raised and the required output temperature lowered, the efficiency can be raised, and achieving this becomes a challenge to both engineer and architect. The heat pump has to work harder in winter and loses efficiency when the outside temperature is much lower than that of the heating system. This fact puts air source heat pumps at an apparent disadvantage in the winter, and has resulted in much promotion of ground source heat pumps, which can rely on a constant input temperature throughout the year of about five or six degrees. Laying the undersoil pipework, at an optimum depth of two metres or more, is, however, inordinately expensive and disruptive, and by no means all houses have a garden that can be used for this purpose. In any event, whatever advantage they may have over air source systems can be cancelled out by preheating the input air. The ways in which this can be done will be looked at below. Again this is as much an architectural as an engineering challenge.

As regards lowering the temperature of the output, since most of it will be required for space heating, the answer lies in coupling the heat pump to a low temperature heating system, which typically operates at 40°-50°, as against 70°-80° for normal central heating. Underfloor heating operates within the first range, as do other low temperature systems which use specially designed radiators. A low temperature heating system can increase the COP of a heat pump from an average of about five to as much as seven or eight. The significance of these specimen figures is that for every kilowatthour of domestically generated electricity used to drive the system the heat output will be multiplied by that amount. It is as close to something for nothing as anyone might desire.

As regards cost-efficiency to the customer, an average air source heat pump may be expected to cost about £5,000 installed, against which the Renewable Heat Initiative in 2012 proposes a subsidy of £850 towards the purchase. The value of heat pumps is so great that it would probably be worthwhile for the government to give one free to every home. In the end the householder foots the bill as taxpayer anyway.

### 9.3 Energy Concentration

Energy can be concentrated in time or in space, or both together, the laser and capacitor being obvious examples. A certain density of energy is required to enable work to be done, but the energy itself comes from the tendency of a system to find a ground state, as explained earlier in the RTE principle. A system which is far from equilibrium contains more potential energy and requires less trigger energy to release it.

Energy concentration can be passive or active. The most obvious passive methods are the solar dish and glass or plastic lens; the most obvious active methods are compression through an electrically driven pump and using electrical resistance to heat water, thus increasing its thermal pressure. Heat pipes may be seen as devices for transporting or concentrating heat. This is, however, a field without clear boundaries.

## 9.4 Energy Conversion

The shape of this problem has appeared above in comparing photovoltaic with solar thermal panels. With PV panels it is a straightforward process of converting direct to alternating electric current, using an inverter that creates a parasitic loss of electricity of about 8%. It is perhaps worth noting that AC current was universally adopted largely to avoid transmission problems over long distances, and early installations that served a locality, before the Grid came into being, invariably used DC. When the emphasis shifts to microgeneration and long distance transmission is not a factor to be considered, it is natural to ask whether there is any advantage to be gained in converting DC output from solar sources and battery to AC to make it grid-compatible. It would be possible to split the total output into AC for the Grid and DC for the house, after adapting household appliances for DC. Alternatively, one could have both an AC and DC wiring system in the house, since some appliances run more efficiently on DC and the significant energy loss from inverters could be avoided. Overall, however, it does not seem worthwhile at this point to become involved in a detailed assessment. That said, off-grid homes and caravans rely solely on a DC system, and there may be benefit in the future in undertaking an in depth study, starting from their experience.

The more immediate problem is how to convert radiant heat from the sun, concentrated by a solar dish, trough or lens, to electricity. Whatever system may be shown to be feasible, it will have to contain an alternator and an engine of some kind to drive it. On the meso- and macro-scale the answer has been found in a composite system of solar dish or trough connected to a Stirling engine which drives a generator. A typical configuration is shown in outline in this diagram.

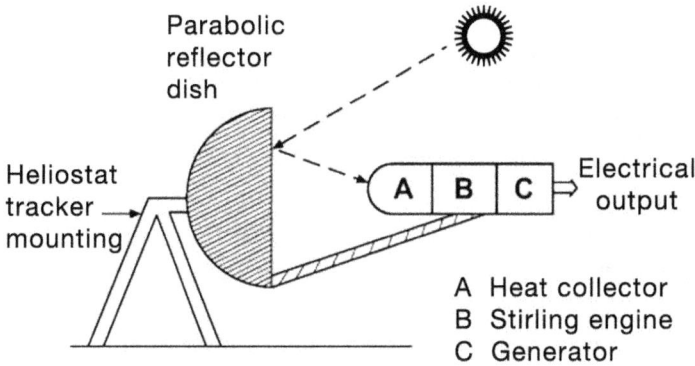

Fig. 8    Solar dish Stirling generator

Although there are already huge "farms" of hundreds of such dishes, each unit typically three metres in diameter, there has not been a similar initiative applied to scaling the device down, though there is no reason in principle why a domestic size version should not function just as well. An overview article in the *Journal of Solar Energy Engineering* makes the point that the solar dish has "the highest efficiently of any solar power generation system by converting nearly 30% of direct-normal incident solar radiation into electricity",[4] and predicts that a model currently costing $10,000 and suitable for distributed generation (i.e., mesogeneration) will eventually come down to $3,000.

A domestic sized installation at this price would be very efficient and cost-effective, and ripe for development. A model with a one metre diameter dish is available commercially at just over £300, but not advertised as a serious augmentation of normal electricity supply. The advertiser does not give a rated output, no doubt because it would function at its peak only when the dish is pointed into an overhead sun. One could estimate maximum output of electricity at perhaps 250 watts at noon on a summer day in the UK, and proportionately lower as the sun rose or set, but maximized by using a heliostat mounting to track the sun. Due to the lower inclination of the sun during the winter months, one might obtain a maximum of 50-100 watts, but this is no more than a first rough estimate. From an economical point of view mounting the dish on a dual helio-stat mechanism would obviously add to the capital cost but savings could probably be made by replacing the dish with a plastic Fresnel lens of roughly the same area. On a good day the temperature at the focal point of dish or lens of an area of about 1.5 square metres is considerable, sufficient to melt an aluminium can. Using a heat diffuser this energy can be directed to drive a Stirling engine.

The weakness of the winter sun (the so-called cosine trap) is where all microgeneration systems come to grief. Just when the house needs more energy for heat, solar energy diminishes. This should prompt the planner to turn attention to solar thermal panels, since they are very efficient collectors and pick up a great deal of their energy from ambient light, before converting it to hot water, thus functioning quite well when the sun is either hazy or hidden. How to convert the energy in hot water into electrical power becomes a critical part of a microgeneration strategy and, fortunately, the answer is at hand in the form of the organic Rankine engine, usually abbreviated to ORC, for organic Rankine cycle. The name is somewhat misleading, since there is nothing particularly green or ecological about the engine. At its heart is the same sort of energy generating loop illustrated above in Fig. 8, *Closed loop state-change energy generation*, in which the evaporation/expansion of a working fluid which boils below the boiling point of water is used to drive machinery. The magic of the ORC is that it can power an engine, and thus generate electricity, from low temperature input heat, although its efficiency increases rapidly with the rise in temperature and the decrease in temperature of the "cold sink" condenser. It would therefore make engineering sense to incorporate a cooling component into the circuit. The diagram illustrates very simply the principle behind the ORC.

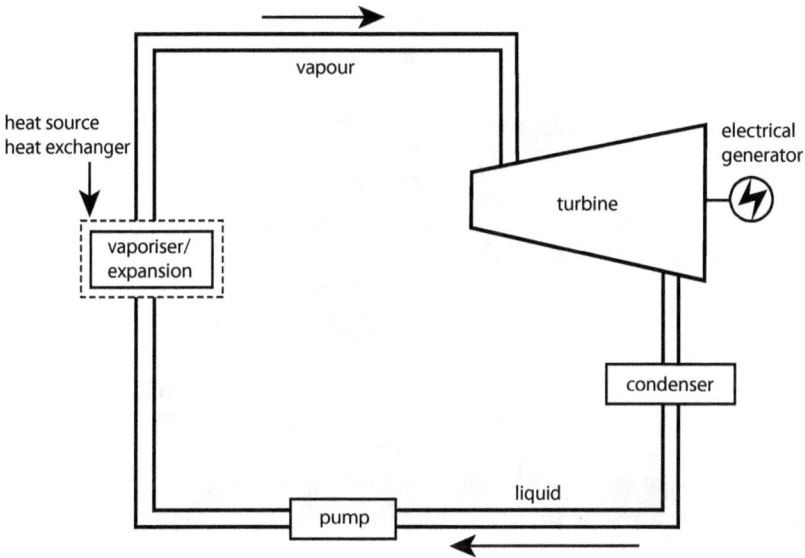

Fig. 9   Simplified schematic of the Organic Rankine Cycle

There are several companies offering ORC's that generate electricity from industrial waste heat, but application of the same principles to domestic scale energy management is a largely neglected area.

## 9.5   Energy Conservation
The obvious place to start with energy conservation is to minimize the heat that escapes from the house by way of conduction, convection and radiation, as shown in this diagram.

Roof 25%

Walls
35%

Floor 15%          Doors and windows 25%

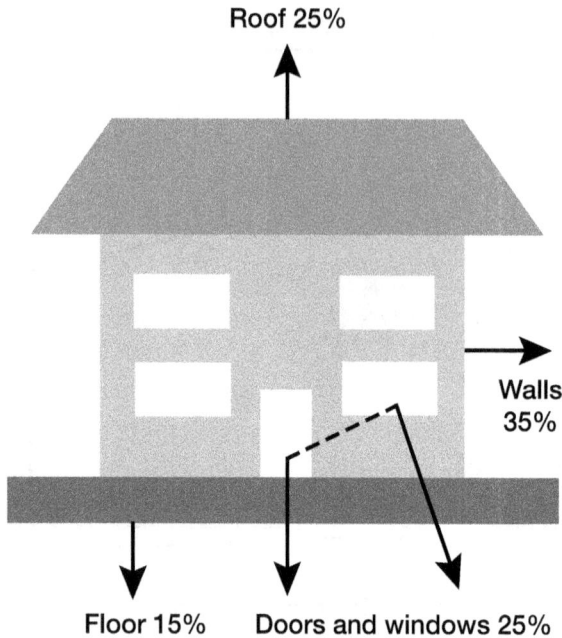

Fig. 10 Approximate heat loss from average home

Along with this, heat loss from hot water cisterns or piping in the house should be minimized by lagging but to a higher specification than has been normal. Each of the areas where heat is lost calls for a different kind of insulation, as well as for design features to eliminate bridging losses and draughts from doors and windows. The latter factor is, however, ambiguous, since in most houses draughts provide an unplanned method of ventilation. A properly planned house today should have an integrated ventilation system which uses ducting and a fan to heat incoming cold fresh air from warm stale air.

### 9.51 Insulation

The average amount of energy in kilowatthours used in the UK home on a typical winter day breaks down (to within 10 -20%) as follows:

| | |
|---|---|
| Heating | 12 |
| Hot water | 3 |
| Refrigeration | 0.8 |
| Lighting | 0.8 |
| Washing/drying | 1.2 |
| Cooking | 2.5 |
| Computer/misc. | 1 |

It is clear from these figures that if the heating requirements of a household could be reduced by, say, ninety per cent - which is less than the mark set by the Passivhaus - the national energy requirement would be cut so drastically that the immediate energy problem would almost be solved at a stroke. The assertion needs qualification, for most space heating comes from gas, rather than electricity, and "energy requirement" includes the needs of industry and transport. Broadly, domestic, industrial and transport each accounts for a third of the total national need, so excluding the last item, where energy need is mainly for petroleum products, if all the houses in the country could be insulated to anywhere near the Passivhaus standard, gas and electricity needs would be cut by about a quarter. How feasible would that be if a coordinated policy could be mounted?

There are some 24.5 million dwellings in the UK, of which about two thirds are of cavity wall construction and of these, according to the National Insulation Association, some nine million are without insulation, as well as fifteen million lofts. Since approximately 60% of the heat of a house is lost through the walls and roof, this is clearly the place to begin. The cost of insulating the average home is £400-700, and, as noted earlier, tax free saving could easily be £200 per year, this is an all-round winner.

A new-build E-plus house is able to set a much higher standard of thermal insulation. Like the Passivhaus, it is conceived ideally as a perfectly insulated box, and from the point of view of heat retention, it is logical to build direct onto a thermally insulated concrete raft. The raft itself can be visualized as a sandwich of reinforced concrete, damp insulation membrane, a thick layer of polystyrene or polyurethane type foam (available in board form with reflective foil) and a protective covering of board on which carpet or other surface finish can be laid. Ahead of actual testing, and with cost-effectiveness always in mind, it would be reasonable to specify about 300mm of concrete and 400mm of extruded polystyrene or polyisocyanurate. The tendency of this overall thickness to compress under weight would call for some form of reinforcement, which would create risk of thermal bridging, but this would not present insuperable difficulties.

In a newbuild the shape of the house will play a major part in planning for heat retention, the aim being to have a minimum surface area. The geometrical ideal would be a sphere, but since this is clearly impossible, the closer one can come to a cube, without jeopardising other factors, the better. Long narrow houses would be avoided and while curved surfaces may be theoretically desirable, they are rarely practicable, if only because of the need for flat windows. The Passivhaus aims for 100% thermal efficiency and Passivhaus certification is given to buildings which consume less than 15 kilowatthours of energy per square metre per year. Emphasis is on a perfectly sealed space, as far as possible, with detailed attention given to external and internal insulation, triple glazed and energy-reflecting windows (E-glass), and elimination of thermal bridging. Efficiency on one scale is measured in the number of winter days when the house can be kept at a comfortable temperature using only the body heat of its occupants (about 100 watts per adult) and waste heat from lighting, refrigeration, etc. That is not always enough, especially in the larger house, but gives an idea of how efficient it is in conserving energy and why it acts as a "gold standard" for the E-plus house to emulate. The Passivhaus concept does not, however, extend to energy generation, and the standard design assumes that energy will come in the normal way from a fossil fuel source (gas or electricity) or from a biomass stove.

## Windows

Triple glazing should be taken as standard. It has long been the norm in Scandinavian countries, where temperature can drop to -20° or lower for extended periods, but there is debate as to its cost-efficiency in the more moderate climate of the UK. However, the terms of the debate are changing as the cost differential between double and triple glazing is narrowing and the efficiency of both is raised through advances in glass technology, frame design and materials that reduce bridging loss. Not considered here is the vacuum-glazed window panel, which operates on the principle of the Dewar or vacuum flask. Guardian Industries in America unveiled in 2006 a super-efficient window, which is being promoted by the American Department of Energy, with an 8 mm gap and an insulating value three to five times better than normal double glazing.[5] There are, however, ongoing difficulties in maintaining the vacuum and a constant gap, since the vacuum pulls the panes together, and these factors together with a much higher cost probably account for the fact that the vacuum-glazed window has not taken off despite early publicity. As regards more conventional, argon filled double and triple glazing, there is a dramatic gain in going from single glazing to top specification double glazing, namely 5.7U to 1.5U and triple glazing takes that down thereafter to 0.8U. While the last figure may seem worth paying for, it is still not very efficient when set against the standard of 0.3U for wall insulation currently demanded by Building Regulations. Even triple glazing lets out a substantial amount of heat. However, as well as the actual loss of heat involved, there is a comfort factor to be taken into account, and triple glazing offers a distinct advantage here over double glazing. Passivhaus data shows that on a freezing cold day, when an internal temperature of 21° is required, the temperature next to a single glazed window can be as low as 1°, next to a double glazed window 11° but as high as 18° next to a triple glazed window, making the room much more comfortable and minimizing convection currents - i.e., internal drafts. On balance, triple glazing with insulated frame seems to be the best choice, especially since the price is constantly reducing, as specification rises. Heavy curtains that can be drawn at night time will act as an extra thermal barrier.

A word should be said here about the three different ways in which thermal efficiency is measured - four, if one includes the Tog rating for duvets. In the UK it is given as a U-value, sometimes as a

K-value and in the US as an R-value. Where the R-value measures thermal resistance, so that a high figure is desirable, the U-value measures how much heat will pass through an object made up of a particular material, strictly how much heat per unit time. The K-value is a composite obtained by adding the thermal resistance of different components in a so-called "thermal object", such as a wall, made up, for instance, of brick, insulating board and plaster. All this can make for quite complicated calculations, but the main criterion for decision is that a low U-value is desirable. If a wall has a U-value of 1, then 1 square metre of it will allow 1 watt of energy to pass through when there is a temperature difference of one degree (i.e., thermal pressure) between the outside and inside. Strictly, the rate at which heat flows through a particular material is the prime criterion, for in theory a metre of brick will function as well as 5 cm of polystyrene in equalizing thermal pressure between the outside and internal temperature. This general observation has several implications, for a moment's thought will reveal that much heat from the house will go towards heating the brick, and hence high-efficiency internal insulation should be the ideal, to prevent heat entering the brick and dissipating. Not so obviously, any insulation material itself absorbs heat, albeit slowly, and thus maximum efficiency is a function of thinness of material. This broad principle needs to be applied also in selecting the material for the internal walls of a house, which one would not normally think of as requiring insulation. More will be said in the following chapter, but in anticipation it should be pointed out that an energy efficient house should incorporate internal walls designed to absorb minimal heat. Strangely, this appears to be a point entirely neglected in the Passivhaus requirements and, more generally, in the architectural literature of energy-efficient buildings.

Insulation is a science in itself, and offers the architect a wide range of options.[6] Insulating material comes in rolls, panels, batts and loose infill, but in practice a choice reduces to a trade off between thinness and cost. The most efficient insulator is aerogel, a fairly recent development, which is essentially a plastic-like sheet full of minute bubbles of a vacuum or gas, which block the passage of convected, conducted and even radiated heat.[7] A typical one centimetre thick aerogel panel has roughly the same insulating power as seven or eight centimetres of polystyrene foam, one of the best materials on the market. The biggest drawback is price, but as development has proceeded over the past five years, that has been dropping dramatically. The rate of development has, in fact, been so rapid that trade information on aerogels is likely to be out of date within months of publication. The great advantage of aerogel-based products being their thinness, they have particular value when it comes to retrofitting insulation, where space is often at a premium. In such situations a compromise between cost and efficiency may sometimes be found in using flexible thermal lining, about 8 mm thick, essentially thick paper on a polystyrene backing. This can be applied like wallpaper, and although not so thermally efficient as aerogel, or even self-insulated boards, is often the only realistic solution in a small room. This by no means exhausts the options, and there are well tested brands of insulated dry lining that can be widely applied for internal wall insulation. Reflective foil can be used on its own to minimize radiated heat or as part of a sandwich with other flexible insulation. The best examples of this kind of thermal barrier will give an effectiveness of 0.18U for a thickness of about 25mm, which is currently required for walls by Building Regulations. The Regulations themselves are a fairly recent introduction, are being continually upgraded and are now of some complexity. In the space of about two years the relevant Part L requirement has changed, for example, from 0.25U to 0.13U for floors and 0.26U to 0.18U for walls, representing overall a halving of the acceptable heat loss. The introduction of yet more demands on architects and builders has had the desirable effect of raising awareness of the importance of insulation. Ideally, a two or three day course would be required to become familiar with all the different types and their usefulness, since there are now so many options, ranging through mineral wool, fibreglass, polyurethane, polystyrene, sheep wool and reflective foil. Most of these can be used as filling for cavity walls, and solid walls can be insulated with structurally insulated panels (SIP's), complemented externally with an insulating render.

The greatest source of heat loss in the average house is usually the hot water cistern, and hence the place where the most efficient insulation should be used. It has been best practice for some years to encase the hot water tank in a box filled with polystyrene beads, thus creating a very efficient thermal barrier of some 300 mm at minimum cost. While this is very cheap solution, an alternative using an aerogel jacket, which frees up space, is now possible for little more expense.

### 9.52 Energy Storage

Energy conservation in the form of storage raises interesting problems at all levels of the system and involves several criteria. A 2014 report by the Institution of Mechanical Engineers, entitled *Energy Storage: The Missing Link in the UK's Energy Commitments* provides a good and partially quantified account of the current options and areas of development, which supplements the information in this section, but is oriented towards central generation.[8] As regards short-term storage in a microsystem, the most obvious and initial need is for each house to store enough energy during the day to meet a 24 hour requirement, timeshifting supply from periods when there may be a surplus. The daytime demand curve goes from a minimum about 10 pm, which is when households start to switch off the television and heating and go to bed, and rises rapidly at about 6 am, when the families wake, have showers and boil kettles for breakfast. The breakfast demand peak occurs some six hours before the sun's energy reaches its supply peak. The obvious solution is to use batteries for short term storage, and off-grid systems are available in which batteries are used in conjunction with solar panels and back-up petrol or diesel generators. However, this does not address the needs of the ordinary grid-connected house, which are inseparable from the needs of the Grid. The complexity of the problem can be appreciated when the supply problems of the Grid are taken into account. If several million houses could be equipped so as to generate, say, 120% of their electricity requirements, the Grid would need to organize the surplus and, ideally, store and feed it as required nationally. It can thus be seen that, as regards energy storage, a K-gen/E-plus system will eventually call for some kind of symbiosis between the individual house and the Grid.

It is a problem that needs to be walked around, for if the ordinary house could incorporate medium or long term storage to cope with a period of perhaps eight weeks in winter, when there may be a dozen sunless or windless days in a row, the Grid would be relieved of a critical problem. Long term storage on the macro-scale is practically impossible at the present time, other than by pumped hydro, but if it could be achieved on the micro-scale, the national problem would be solved. Storage is, indeed, a vital component in the E-plus system, since without it the Grid would always need to have almost a full complement of conventional installations on standby. As will be shown below, short and long term storage are now well within reach both technically and economically.

Storage for space heating demand can be solved quite simply by incorporating a tank within a tank containing chemicals with appropriate phase change characteristics, such as Glauber's salt, magnesium hydride, pure graphite or paraffin wax. These would liquefy or dissociate chemically as they were heated, when solar or other energy was available, and release heat as needed on cold days through a thermostatic system. The tank need not be very big, since in a house insulated to 90% of the Passivhaus standard, very little space heating will be required, and storage might take up perhaps $2m^3$. Heat loss would need to be no more than about 5% per day, calling for exceptional insulation. A fan mechanism would be needed to circulate the warm air when required and some specific knowledge of the chemistry of phase change materials.

The basic principle involved here is merely an extension of systems tested out fifty years ago on houses which used hot rocks in the roof space or basement to store the excess heat of the sun. The same principle is currently being pursued by one British company, Isentropic, [9] but enhanced by heat pump technology and on a large scale, with Grid compatibility in mind. This is achieved by a simple arrangement based, to quote from their literature, on "two large containers of mineral particulates [gravel] which use electricity to pump heat from one vessel to the other, resulting in the first container cooling to around -160°C and the second container warming to around 500°C." The containers are elsewhere called "silos" and run to about 350m³, which is about the size of an average house. The specially designed heat pump machine is so configured as to operate as an engine, which drives an electric generator. It is obvious why a large scale installation has been chosen, for the differential between these two temperatures (and thus the thermal pressure) is exceptional. Notwithstanding, the system is offered as "modular and scalable". While it is probably uneconomical to scale it down to single dwelling usage, there is no reason why it cannot be designed on the meso-scale to service a dozen houses. This would be desirable for several reasons, not least because it would bring electrical generation under the control of the user and make it cheaper by cutting out the supplier's profit. There is no information available at this point about future cost-efficiency.

While awaiting the results of research into this method of storage, the immediate challenge facing the E-plus house is storage of electricity, rather than heat, and that leaves the planner with two alternatives, namely, electrochemical and electromechanical storage devices.

Electrochemical storage, more usually referred to as batteries, is one of the fastest developing fields in energy science. It is hardly possible to summarise the important criteria of different types of battery - storage capacity, rechargeability, acquisition and maintenance cost - for purposes of comparison. In general, however, it may be said that while rechargeable lead acid batteries have not changed significantly in a century, great advances have been made in lithium-ion, nickel-cadmium, sodium-sulphur, lithium-air, zinc-air, solid electrolyte lithium batteries and others. Lithium based batteries have been intensively researched and used in electric vehicles, have halved in price in five years and have much to offer for domestic use in terms of capacity, speed of recharge and duration of charge. However, it is not clear how far the world's supply of lithium would be able to fulfil all future needs. One expert in the field offers the opinion that "analysis of lithium's geological resource base shows that there is insufficient lithium available in the earth's crust to sustain electric vehicle manufacture in the volumes required .... Depletion rates would exceed current oil depletion rates and would switch dependency from one diminishing resource to another." Significantly, he adds, "Concentration of supply would create new geopolitical tensions, not reduce them."[10] Sodium-sulphur batteries are highly efficient and with no problem regarding future supplies of the basic materials. They are, however, difficult to manufacture and maintain, since sodium is so corrosive. The future may lie in zinc-air batteries. ReVolt, a Swiss company claims that their version can store three times the energy of a lithium-ion battery, by volume, while costing only a half as much. The overarching question is, when the overall best type of battery has emerged from present competition, what role will it, or can it, play in a K-gen/E-plus system?

Electromechanical storage systems, more commonly known as flywheels, have been around for some time. Anatoliy Ufimstsev constructed a wind-powered generator in 1931 combined with flywheel, which supplied electricity to his workshop and several homes in St Petersburg. However, the flywheel was seen essentially as a buffering mechanism, not for long term energy storage. The change in approach has come about only in the last twenty years or so and as result of three design and technological breakthroughs. It would hardly be too much to call them the magical answer to the of problem energy storage and thus ultimately to the global and national energy crises outlined earlier. It is easy in principle to power up a flywheel by electric motor and draw upon the energy of rotation to drive a generator when demand arises. The problem in the past has been that frictional losses depleted the stored power too rapidly to make the flywheel of much use. Now those losses have been almost entirely eliminated. With the invention of the Halbach array in magnet technology friction from the bearing has reduced almost to nil by magnetic levitation.[11] Maglev bearings are not only virtually friction free but avoid the stresses which tend to destroy ball bearings through vibration, though they bring with them other technical problems, particularly with balancing. The other critical advance has been made possible by encasing the whole assembly in a vacuum enclosure, thus eliminating the considerable friction from surrounding air. The third technological breakthrough has been in the use of carbon fibre composites to replace the solid steel of traditional flywheels. Since energy is proportional to the square of tip speed, a rapidly rotating wheel generates extreme radial stress, and thus it is more important to make the flywheel not of the densest material but of the highest specific strength. New composites in a vacuum environment are capable of rotational speeds of over 35,000 rpm.[12] Given that stored energy is proportional to mass multiplied by the square of angular momentum, it is easy to see that this can represent a great deal of energy even in a small flywheel.

The new generation of flywheels appears to be ideally suited for small scale storage needs, since the sheer amount of dynamic energy contained in larger versions makes balancing very powerful gyroscopic forces more critical. However, medium capacity versions may prove the best all round compromise, and this remains to be tested out. Mesoscale storage with either a single larger unit or several smaller linked units may prove to be the most convenient and cost-effective. Mesogeneration would also offer a compelling advantage when it comes to siting the unit below ground, as is desirable for safety reasons. A single medium-size flywheel would certainly be more economical than siting many models in cellars and gardens. Thinking along these lines leads towards a conclusion that a future Grid may be designed around an electromechanical storage system on the mesoscale, as illustrated in figure 10 below. In an ideal system of this kind, the many thousands of flywheel modules dispersed across the country would serve a dual control function, as a permanent storage for the Grid, to be tapped like a hydroelectric installation and, equally, for the individual home to tap into when there was insufficient supply from sun or wind energy. The shape of a fairly complex control system is now coming into view.

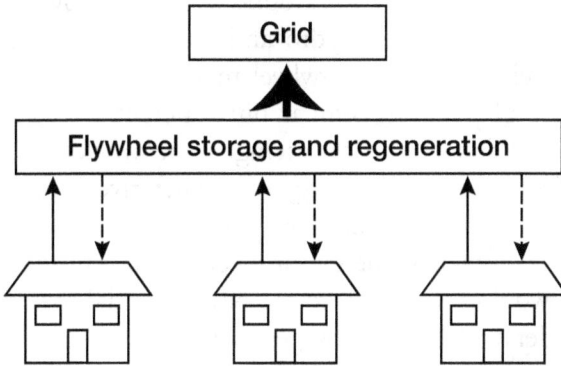

Fig. 11 Flywheel storage subsystem

The above information is only a brief summary of the state of the art in flywheel technology, which itself is a rapidly moving front, but should be sufficient to indicate the revolution in energy management that has been set in train with breakthroughs in flywheel design. With virtually no energy loss, unlimited service life and minimal need for maintenance, it is capable of filling not only domestic, industrial and Grid storage requirements, but also the widespread need for Uninterruptible Power Supply (UPS) for hospitals, IT centres and other socially vital places, which is currently met by petrol or diesel generators and lead-acid batteries. Fully charged, a modern flywheel linked to a generator will suffer less than two per cent of idling loss in a month, and there are experimental examples that have been running for two years without significant loss of power. Hence a domestic size installation which could take a house through a worst case of two almost sunless and windless months in the winter would need a capacity in the region of perhaps 200 kWh. That would call for a flywheel complex (variable speed driver, flywheel, generator, load stabilizer and online monitoring and display) measuring no more than 1 x 1 x 2 metres, and weighing much less than the equivalent volume of lead-acid batteries. Cost-effectiveness is not easy to calculate, since no other device has this function and it is hardly possible to assess the value of storing energy that would otherwise be wasted. The Velkess system claims to "deliver distributed and highly scalable storage for around $1,333 a kilowatt, making it competitive with pumped hydro and compressed air."[13]. There are several well established flywheel manufacturers in the US and in Germany, all pursuing similar but different tracks. Some concentrate on UPS, some on car racing, some on electrical transport systems. The latter use the energy of regenerative braking to power up trackside flywheel installations, which is then fed back to the tram as electricity when energy for acceleration is needed. This general model would seem to have the best potential for developing a specific model dedicated to a K-gen system.

### 9.53  Waste Heat Retrieval

There are two ways in which waste heat can be put to use in conserving energy on a domestic scale. A heating, ventilation, air-conditioning system (HVAC) which is standard in Passivhaus design, conserves energy by prewarming fresh air, using heat taken from extracted stale air. The fresh air inlet is normally in the roof and heat exchanger in the roof space. An efficient system obviously depends upon the house being sealed, with no draughts from windows or doors. While it saves on heating costs, there is ongoing electricity consumption arising from the fan which pumps the air around. However, this is a very low pressure circulating system, probably requiring no more than 50 watts per hour, perhaps a third of the heat energy saved. There are already off-the-shelf systems available.

The second kind of heat retrieval system makes use of waste hot water, from baths, showers and dishwashers, which is really an architectural problem and will be looked at in the following chapter. All that will be noted here is that if the house is approached intuitively as an integrated system, it is natural to see the heat from waste water as a way to preheat incoming air in an air source heat pump.

### 9.54  Voltage Optimization

A significant amount of electrical energy is wasted because the voltage of the electricity as delivered is higher than is needed for the appliances it serves, both in commerce and in the house and, additionally, supply may fluctuate by anything up to five per cent. The difference between supply and requirement voltage may be substantial, with supply typically about 240 volts and appliances designed to operate at 220 volts or lower. Where the electricity is needed for heating purposes there is nothing to be gained by decreasing the voltage, since all that happens is that the heater will run longer to generate the same amount of heat. Most other appliances, however, will theoretically save about ten per cent on their electricity requirement, which is the claim made by the only company selling optimization devices in the UK. VPhase, currently listed on the AIM market and endorsed by Ofgem since 2009, offers a "fit and forget" portable device for about £300, which should in theory pay for itself in four or five years. Because the excess and unusable voltage would normally be converted to heat, the company can claim that use of their voltage optimizer protects wiring and appliances from long term heat damage.

### 9.6   Control

The K-gen/E-plus ideal is a low or no maintenance system that will adjust automatically to the various needs of the house and the Grid. This calls for several kinds of control, notably linking, sensing, modulation, switching and dumping in the overall cause of homeostasis. Not much more can be said, or needs to be said, until the actual components have been specified. A wide array of control devices has come into existence with the invention of the microchip and computer and there is no reason to think that what might be needed in the future cannot be invented. Within this context, three particular kinds of control processes are perhaps worth mention as examples of what can be done. The first, already mentioned, is incorporation into computers and televisions of "standby" cut-off switches which would act automatically after a given delay of perhaps ten minutes. The Energy Saving Trust estimates that unnecessary time spent in standby costs the UK "up to £1.3 billion a year," but seems to favour more education for the public in this matter, as against government action, which admittedly would not affect existing equipment. A valuable interim step towards regulation would be to promote product labelling. A second example worth citing is "ripple control", already used in South Africa, which enables the grid to send high frequency pulses through the mains electricity to switch off and restart components which are not time-critical, such as immersion heaters or washing machines, thus enabling peak requirements to be reduced in real time. No doubt this is being seriously discussed by the appropriate authorities in the UK, for the potential advantage to the Grid could be very significant. Demand in the UK varies between 20 and 60 gigawatts, and if all the nineteen million immersion heaters were switched on at the same time, they would draw 55 GW. If only a twentieth of them were to be switched on at any one time, a reduction of peak load by 3 GW could be achieved by this automatic switch-off method. The third well established method is dynamic demand control, which is similarly used to manage consumption of electricity in response to available supply. In this case, the grid can reduce demand at peak times more sensitively, by slightly reducing the voltage for individual items in the house, such as air-conditioning or clothes dryers, each of which will have its own control box. In being able to selectively control demand in this way, the electricity supplier is able to make disproportionate economies in the number of stand-by installations that would otherwise be needed.

### 9.7 Outline of the K-gen System

Summarising the information in this chapter, the defining elements in a system of the K-gen type are solar thermal panels and a novel wind turbine to collect energy, a heat pump to harness state change energy, a Fresnel lens to increase energy density, an organic Rankine engine to convert low grade heat to mechanical torque and a flywheel coupled to a generator to store and release energy as electricity. The diagram below displays this as an energy flow, starting with the input of solar thermal energy and going through the various processes of enhancement, storage and conversion to delivery of energy as AC current. Ideally the flow chart would quantify energy input at each stage and subtract parasitic losses, but the storage function complicates a simple flow model and since systemic gain and loss occur at every juncture in the flow, and weather conditions are only predictable within broad limits, reliable figures can only come from actually building and testing the system.

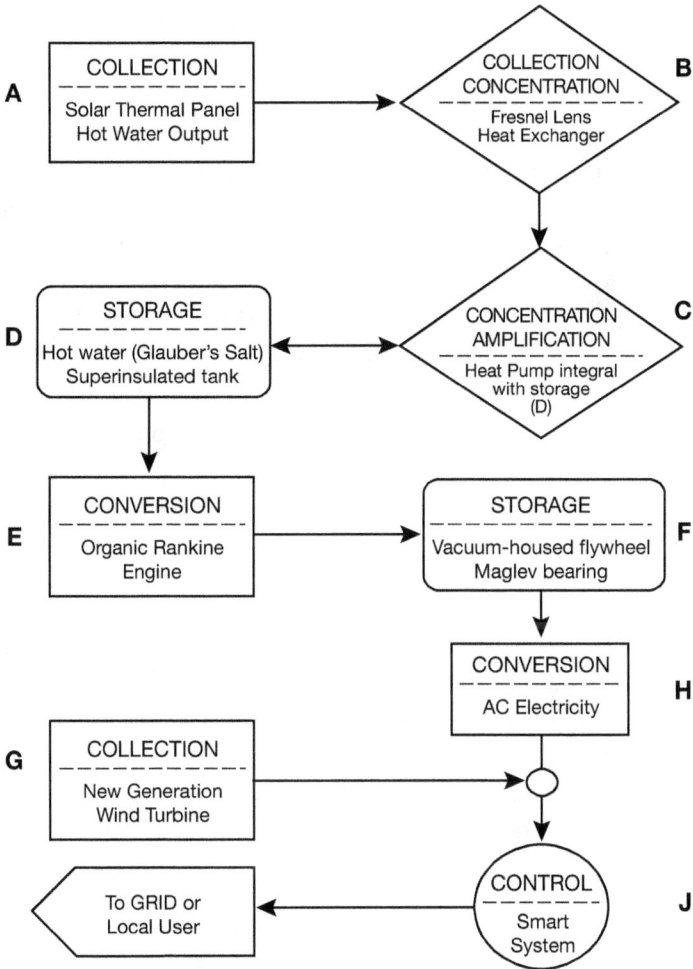

Fig. 12 Schematic for a K-gen system

The heuristic nature of the research has been noted earlier, but should be emphasized, since it is customary to construct detailed mathematical or computer models of projects at concept stage to assess their feasibility before embarking on the actual work of constructing them. In the present instance this is possible only to a limited degree. If the concept can be proved, it will be a "game-changer" in the most fundamental sense, for a workable system will displace conventional forms of energy generation and will usher in a new industrial revolution, as did the steam engine when it made wood burning and horse power obsolete. It is hard initially to take in how high the stakes are and claims for the system can carry little weight until it has been proved, or otherwise, to work as the logic predicts. A functioning system will depend on each of the subsystems functioning as in theory, but their combined performance creates variables on top of variables, and all within the daily and seasonal variability of the weather. Speculation returns then to the two premises on which the whole project rests, that there is ample energy for the world's needs in the sun, wind and the untapped energy of state change and there is ample engineering expertise in manipulating energy to do this, once it is all brought together. The real research challenge will be bringing it together.

## 9.8   Questions of Scale

The original research for the book was done on the assumption that the system would be designed to fulfil at least the energy needs of the individual house. That aim, the E-plus system, still holds but it soon became clear that significant efficiencies would be gained by approaching the problem of clean energy generation as two separate but linked systems, namely, energy collection/conversion and building design. The latter would be aimed largely, but not entirely, to minimizing energy use through superinsulation and low temperature heating. Isolating the system for generating electricity as far as possible from architectural issues clarifies the engineering requirements, enabling them to be treated separately as the K-gen system. Freed of the constraints of modelling on the scale of a single dwelling, the engineering problem was simplified, but also changed shape. In effect, it became two problems, namely, designing a single package solution to be sold to the individual householder and an alternative method of mesogeneration, answering to the needs of the Grid. Indeed, it will be apparent from the block flow chart above that it can be realised as macrogeneration simply by scaling up the component parts.

In this situation it would make sense to conduct research on the smallest, quickest and most economical scale, making adjustments thereafter from the experimental data as it emerges, not least regarding weather patterns that could be relied upon. Two factors biasing the system towards meso-scale are the desirability of erecting wind turbines at more than rooftop height to maximize wind energy and, ideally, locating the flywheel below ground for safety reasons. However, neither of these factors is an absolute requirement.

Once the concept has been proved, expanding the model to macroscale can be treated as a separate problem. While commercial exploitation would initially seem to offer cheaper electricity, it must be remembered that a very significant value of the K-gen system is that it provides energy in the form of hot water directly and cheaply to the household, which would not be the case with macrogeneration. Also, the larger the scale envisaged, the more need is created for infrastructure in the form of transmission lines, transformers and breakers and this would be a hidden cost.

These conclusions must of necessity be general, for even if a K-gen prototype were to emerge as the successful outcome of a research programme, it would doubtless be as primitive and undeveloped as Newcomen's first steam piston engine or the Wright brothers' "Flyer" and a lengthy period of improvement would be required to bring out all its potential. It took 150 years to go from Newcomen's lumbering and monstrously fuel-greedy steam engine to the elegance, speed and fuel efficiency of Gresley's and Stanier's racehorses, the immense pulling power of the Union Pacific's "Big Boy" shire horse and Chapelon's almost philosophical concept of the locomotive system. The airplane's development was shorter, but even more astonishing. Only forty years after the Flyer wobbled a few hundred metres at rooftop height, the jet-engined plane was breaking the sound barrier. Once the principle had been proven, a flood of incremental improvements followed, and there is no obvious reason why the principles underlying the K-gen system, once realised in a working model, should not provide the same kind of stimulus to a new wave of engineering design.

## 9.9 Putting in the Numbers

Feasibility will depend critically on the cost per unit of electricity produced in relation to the all-in cost of the K-gen unit, including maintenance and replacement costs. Since the "fuel" is effectively free, when all development, manufacturing, maintenance and replacement costs are factored in, it would not be unreasonable to expect the long run unit cost of delivered electricity to be considerably lower than the current cost of electricity from the Grid. However, against a background of the planet's urgent need to reduce atmospheric pollution from carbon dioxide and UK's even more urgent need for electricity, one cannot define cost-efficiency in the narrow, conventional way. A pint of cold water in the desert is almost literally priceless. Even so, there is every reason at this point to think that when fully developed, a K-gen/E-plus system will pay for all the householders' energy and put a substantial amount of disposable income in their pocket, with proportionate effect on the economy in general.

## References

1. Interview in Elbert Hubbard's *Little Journeys to the Homes of the Great* (1910).

2. Precise data for comparison of winter/summer performance of flat plate and evacuated tube collectors can be found on the website of Energy Matters Australia www.energymatters.co.au/renewable-energy/solar-power.

3. The research has been done by the Australian National University and its commercial application has been vested in Wizard Power Pty. Basic information is available on www.wizardpower.com.au

4. Thomas Mancini, "Dish-Stirling Systems: An Overview of Development and Status", *Journal of Solar Energy Engineering* May 2003, vol. 125, Issue 3, pp. 135-152.

5. See *Zero Energy Windows* on the Internet

6. See, e.g., *www.selfbuildinsulation.com. and Building Research Establishment website.*

7. Aerogel specification can be sourced most easily through the manufacturers and suppliers (see, e.g. Proctor insulation, Aspen Aerogels, Kalwall, Vacupor, Thermablok.).

8. Available as download from the Institution's website: www.imeche.org/knowledge/themes/energy/energy-storage

9. See, www.isentropic.co.uk. Isentropic is a member of the European Association for Storage of Energy (EASE), which "ultimately aims to support the transition towards a sustainable, flexible and stable energy system in Europe." This is one of several such bodies (mostly with the same big company names appearing as members) with high aims which may give the impression that they are actively forming policy. The question returns continually, what governmental or international body should organise and drive energy policy in the face of global crisis?

10. William Tahil, "The Trouble with Lithium". Internet entry, posted December 2006. Tahil is the research director for Meridian International Research.

11. In a permanent magnet Halbach array (invented in 1985) the field produced by each magnet reinforces the fields of all the other magnets on the active side of the array and cancels them on the other side, creating an intense field. A five-element Halbach array provides more than three times as much force per unit volume as a normal opposed face magnet.

12. The "Flybrid" design, used in Formula One cars after cornering, spins at 60,000 rpm, puts out 80 horsepower for 6 seconds per lap and can shave a critical four tenths of a second off lap time.

13. "Turn up the Juice: New Flywheel Raises Hopes for Energy Storage Breakthrough," *Scientific American*, 10/04/2013.

# Part Four

# The E-plus System

*If we are serious about reducing energy consumption from buildings, we need to add an energy sixth sense to our everyday lives.*

Joseph Giacomin, Director of the
Human Centred Design Institute
Brunel University

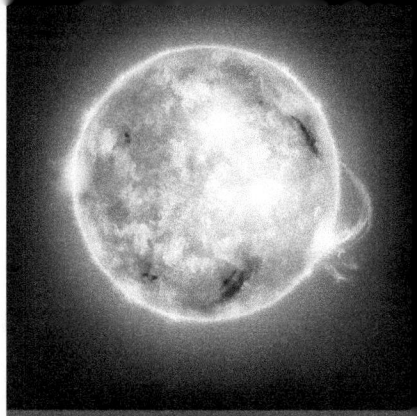

# Chapter 10

## The E-plus System

*My idea of the architect is as a coordinator, whose business it is to unify the various formal, technical, social and economic problems that arise in connection with the building.*

Walter Gropius, The New Architecture and the Bauhaus [1]

### 10.1 E-plus and the Evolution of the House

This chapter will make some general and specific suggestions about the architecture of the E-plus house, a building designed to answer to the future energy needs of the household and the planet. In having this dual focus the E-plus house constitutes a natural - indeed, one might say, an inevitable - future step in the evolution of the house and, to that extent, in the science and art of architecture. The E-plus concept is revolutionary in four clearly defined respects, firstly, in approaching the normal house as its own source of electrical power, secondly, by this means able to contribute significantly to national energy needs, thirdly, in taking the house, or any habited building, to be an energetic system, with all parts coordinated and, fourthly, in seeing it in the widest cultural context as a significant factor in overcoming global warming. Once these principles are accepted, traditional architectural criteria of form and function shift: we are looking at a radical change in the domestic dwelling and a step change in our understanding of what constitutes a "normal house".

The house in relation to its energy requirements can be seen to have evolved from an undesigned shelter, usually a cave with a fire at the entrance, to the nomad's tepee designed with an open top to evacuate the smoke from a central fire and then to more substantial buildings that went with a settled life, initially hardly more than a turf box but developing into brick or stone dwellings with glazed windows and a flued fire or stove. The modern house may be said to have arrived with the advent of indoor plumbing. Only a century ago in Britain a normal worker's house did not have an inside toilet and sometimes not even piped water. Hot water on tap, which we now take for granted, would have been considered almost science fiction, as many today would, no doubt, consider the idea of an energy-generating house.

The transition to a new normality for the ordinary house can be traced to Edwin Chadwick's 1842 report, *The Sanitary Condition of the Labouring Population*, which persuaded the government of the time to build a national water supply and a sewage disposal infrastructure. As late as 1934, well within the lifetime of many today, Sir James Spence's ground-breaking report on child health and sanitation, based on a thousand working class families in Newcastle, recorded that 40% of houses had no bath and 25% still had a shared outside toilet. Personal hygiene was catered for either by public bath-houses in large towns or more usually by a tin bath, filled with kettles of hot water from the coal-fuelled kitchen range, typically used on a Saturday night in preparation for church attendance on Sunday. Until the passage of the Clean Air Acts in the 1950's, heating was invariably by coal fires and domestic central heating was considered a luxury for the rich, whereas now central heating is not worthy of comment. No one today would visit a house today and say, "Oh look, it's got central heating," and in another fifty years, we may anticipate that no one will say in surprise, "Oh look, it generates its own energy and makes a profit selling electricity to the Grid."

In the same way that the reports of Chadwick and Spence initiated a change in domestic architecture, it is logical now to call for a document of equal importance which will show the significance of energy management in design. The present work is no more than an introduction and, hopefully, an incentive to embark on such a project. We need a landmark document, and if any architect should be weeping, like Alexander, for new fields to conquer, it lies before him or her. We are faced now with redesigning the house to deal not only with the wellbeing of the nation, which motivated pioneers in social housing, but the well being of the planet, and the first step must be to prove that such a house is feasible and replicable. Since there is always resistance to the new, simply because it is unfamiliar, and microgeneration (whether or not incorporating the K-gen and E-plus principles) is at the present time unfamiliar and unproven on a large scale, the idea of providing one's own electricity and selling it rather than buying it will seem to go against the natural order of things, for we know only top-down electrical distribution and few in the profession are likely to see this as an architectural challenge. Raising public and professional awareness of the issues will therefore be as necessary in developing the E-plus house as is the design itself. At this point we are looking for pioneers and perhaps initially for a single architect who will feel it as a driving ideal and inspire others, as did Walter Gropius with the Bauhaus ideal.

There is certainly no lack of experimental houses which have set out to utilise solar energy sometimes, but not always, allied to energy saving. Few, however, have gone beyond the concept stage, and none sought a fully integrated solution to the energy challenge. The "Heliotrope" house designed and built by the German architect Rolf Disch in 1994 claimed to generate 400% of its energy needs by turning the whole building to follow the sun, but it cost over £1 million. There have been other house designs on the same principle, such as the *casa giratoria* in Argentina in 1951 and Roland Mösl's cylindrical Gemini house in Austria, which won a Europe-wide innovation award in 1993. A different approach to harnessing solar energy in the domestic residence can be seen in houses designed by Fred Keck and Maria Telkes in America in the 1950's and later by Félix Trombe in France (who will call for more attention). Where the E-plus house differs from all these examples lies in its proactive approach to energy, in collecting and then manipulating it, emphasizing the need for superinsulation in domestic architecture and for innovations in energy storage and utilisation of waste heat and water.

## 10.2   E-plus and a Cultural Shift

If the assumptions of the present work are correct, domestic architecture will take the lead in generating a cultural consciousness such as history normally displays in the architecture of grand houses or public buildings. It will make society conscious of its dependence on energy and on the vital role it plays not only in economics but in all aspects of life. History shows examples too numerous to list of the mutual influence of architectural design and cultural consciousness. One of the most obvious and enduring is the so-called gothic style, which not only expressed the cosmic awareness of the European Middle Ages, but powerfully stimulated the engineering imagination. As against the workaday pragmatism of the Roman arch, mediaeval belief in a two-level universe of heaven and earth, and its cultural significance, was embodied in the pointed arch that soared upwards with breath-taking elegance.

In the last century international architecture on all scales has been dominated by individualism and parametric design in which the computer has, supposedly, enlarged the architect's imagination, though it might be argued that it has anaesthetized it. At least, one can hardly imagine the Taj Mahal, Chartres or the Parthenon emerging from CAD software. That said, with informed usage, CAD does promise useful results in exploring various potentials in the small scale E-plus house. On the grand scale good design has de-emphasized the importance of formal and functional unity and increasingly become associated with "making a statement", often the more bizarre the more applauded. Leaving aside the grandiosity, or worse, of Soviet style architecture, what might perhaps be called generically the consumer-capitalist style has resulted in buildings shaped like gherkins, upturned boats and heaps of broken glass. From the engineering perspective there have certainly been ground-breaking developments of many kinds, but their social value has a deep ambiguity. Daniel Libeskind's trademark style, jagged and asymmetrical, was initially called "unbuildable" and may be accused of creating a challenge to the structural engineer simply for the sake of challenge. It seems, however, to strike a latent chord in a fragmented society, which has not only lost direction but has elevated lack of direction almost to a cultural norm.

The most obvious architectural innovation of our time, in every sense, is seen in the hubristic skyscraper which totally ignores the convenience of the inhabitants or its appalling demands on material and energy. Renzo Piano talks of his London Shard, now dominating the skyline, as a "vertical city", "an adventure", "a fight against gravity" and "a struggle between practicality and spirituality" and draws parallels with the way in which the gothic style integrated all these factors in an earlier age. The reality is, however, that its imposing height is hugely energy-inefficient in its need to transport these new "citizens" up and down, and its tapering profile speaks of profligate waste of valuable floor space as well as energy.

Architectural advance has always come from a dialogue between engineer and artist, a theme developed at length by Sigfried Giedeon in his magisterial *Architecture and the Phenomena of Transition* (1971). He notes in particular the revolution that followed from the completely iron structure of the Eiffel Tower in 1889, although this might actually be seen not as a beginning but as the high point of a revolution that started with the Iron Bridge at Telford in 1781, designed by the now largely forgotten architect Thomas Farnolis Pritchard. In this instance the essential contribution of the construction engineer is easy to see, but from the present perspective of the energy-conscious building, and the E-plus house in particular, the architectural challenge looks very different. The future is seem to depend on the electrical and heating engineer and equally on the chemical engineer when it comes to such demands as more efficient batteries or specifying the best insulating material, on which more will be said below. Until the latter part of the last century insulation had been almost entirely neglected in architectural theory: indeed, it would be hard even to find it mentioned in historical surveys. It should now be at the centre of modern architectural theory and practice and, as will be detailed below, is making its way there.

If the E-plus concept of the energy-managed building turns out to be a significant part of the answer to the world's energy problem, it would have the potential to be the seed for a new kind of architectural revival, comparable perhaps to the Bauhaus movement between the wars in its integrating power but going beyond it, because its inspiration is drawn from a double idealism. Where Gropius's vision was of an architecture fit for a politically and aesthetically renewed society, the vision now arises from the urgent need for a green planet, a sustainable economy and the ideal of a global society drawing almost all its energy from non-polluting sources. The peak of E-plus design would be a house which pushed energy management to the limit but was also, to use a Bauhaus criterion, "a house for an art lover" and "the first letters of a new architectural alphabet".[2]

## 10.3 The Passive Energy House

The energy-conscious house is usually conceived as passive in two respects, in that it focuses on insulation against heat loss and on the passive reception and short term storage of solar heat. The Passivhaus that appears throughout these pages focuses tightly on insulation. It represents the development to an extreme of the concept of a box totally sealed against the intrusion of cold or escape of heat, with special provision made for admitting the fresh air that the residents will need. In hot climates its benefits work the other way, keeping the house cool and reducing the need for air conditioning. Although the Passivhaus is taken to be a trail-blazing idea, its initiator Wolfgang Feist very consciously borrowed from the Conservation House a successful project of the Saskatchewan provincial government in 1977 to counter the rigours of a Canadian prairie winter. The Saskatchewan Conservation House, like the Passivhaus, relied essentially on superinsulation, but also incorporated solar thermal vacuum tubes, and while its subsequent history is difficult to trace, there is no doubt that it raised awareness of the importance of insulation. The principle was taken up by the American Eugene Leger, who claimed that the so-called "Lo-Cal" construction technology (very similar to the Passivhaus) gave dramatic saving on heating bills - $38 for his house, as against about $800 for his neighbours. This point is of particular importance for the E-plus house concept in that it had the effect of making "progressive builders and energy researchers throughout North America sit up and pay attention."[3] Given the inertia and multiple goals of the UK's energy policy and uncoordinated information on energy saving that the ordinary citizen receives from competing companies, a single stark figure like Leger's would be invaluable in stimulating interest and investment.

The first *Passivhäuser* were built in Germany about 25 years ago and the specifications have changed little since then. Only about 30,000 specimens have been built (at 2014), mostly in Germany, despite a great deal of publicity, which leads one to ask why. There is certainly no strong economic case against the Passivhaus, for the additional building costs can be as low as 10% of the whole and can be largely offset against the saving on a central heating system which is not needed. The saving on space heating (effectively tax deductible) is almost 100%. On a mild winter day the body heat of a family, at about 100 watts per person, is sufficient to keep a smallish house warm. Even on the coldest days supplementary heating of only 500 watts (about three incandescent light bulbs) would theoretically be more than enough. One must ask if there are unnoticed reasons, physical or subliminal, to account for slowness in the take-up of the Passivhaus. Physically, the airtightness of the house demands forced air and preheated ventilation, which can have a severe dehumidifying effect in cold weather, and while strategically placed bowls of water and potted plants have been proposed as rather ad hoc answers to this real discomfort, there does not appear to have been a systematic attack on the problem. (One must speak tentatively, as development of the Passivhaus is ongoing.) There may well be some subconscious resistance to the idea of living cut off from nature in such a well-sealed house, particularly perhaps a sense of unease or even deprivation at not being able to hear outdoor sounds. Elimination of traffic noise outside is doubtless to be welcomed, but not at the cost of being unable to listen to birdsong or children at play. There may also be a primeval need in some for the heat and light of a fire or stove. The Latin word *focus* means a hearth, a feature which has functioned from time immemorial as the focal point of the house. A home is not a house, to use Reyner Banham's witty inversion,[4] and designing in the factors that make a house a "place to come home to" calls for architectural wisdom as well as engineering imagination. Without it we end up, to use another of Banham's striking inventions, with an "un-house". While these concerns are of great importance, they may be put aside for the moment. The central fact is that the

Passivhaus is an architectural breakthrough and the basis on which the E-plus concept builds and, hopefully, may even improve.

The ideal of the *Passivhaus Institut*, is a building that achieves as closely as possible zero heat loss. (Again, one must emphasize its ability to insulate against external heat in hot climates.) Certification is awarded only if the house is airtight to the extent of losing less than sixty per cent of its volume of air per hour and maintaining energy depletion at less than 15 kilowatthours per square metre of floor space each year. Certification is not concerned with the aesthetics or layout of the building, so long as these standards are met. Not without interest is the fact that the *Passivhaus Institut* recognized in 2014 the need for something along the lines of the E-plus house in establishing two new categories namely, the "Passive House Plus", for net zero energy performance and the "Passive House Premium" for buildings that generate more energy than they consume. It does not, however, offer specific advice on how to achieve these levels. Builders are instructed to follow only five principles, namely, high performance wall insulation, small triple-glazed windows, complete air-tightness, elimination of thermal bridging (through window and door frames, etc.) and a forced air ventilation system. All of these are incorporated in the E-plus principles. There is neither specification nor recommendation of anything else in the Passivhaus, such as solar panels or heat pumps. So the structural concept is completely passive, with the exception of the ventilation system which calls for a very low consumption electric fan. This is needed to send cold incoming air through a heat exchanger that is warmed by outgoing warm, stale air. The ducting can also incorporate an electrical heating element that can be switched on automatically if the temperature of the house should drop below a pre-set level.

The conventional solar house is equally passive, insofar as it goes no further towards maximizing heat input than increasing the amount of glazing and storing enough in masonry (or the equivalent) in the same way that block heaters are used in off-peak electrical heating systems. Like the Passivhaus, some models use electric fans to circulate the heated air. Heat storage is, however, minimal, no more than enough to tide the house over for a day or perhaps two when the sun is not shining. The efficiency of the passive solar house could, of course, be improved by incorporating features of the Passivhaus. The most passive house, and perhaps the most ingenious, is the "sun tempered house" of Socrates, designed nearly 2,500 years ago and described by his one-time pupil Xenophon almost as a geometrical solution to a thermodynamic problem.

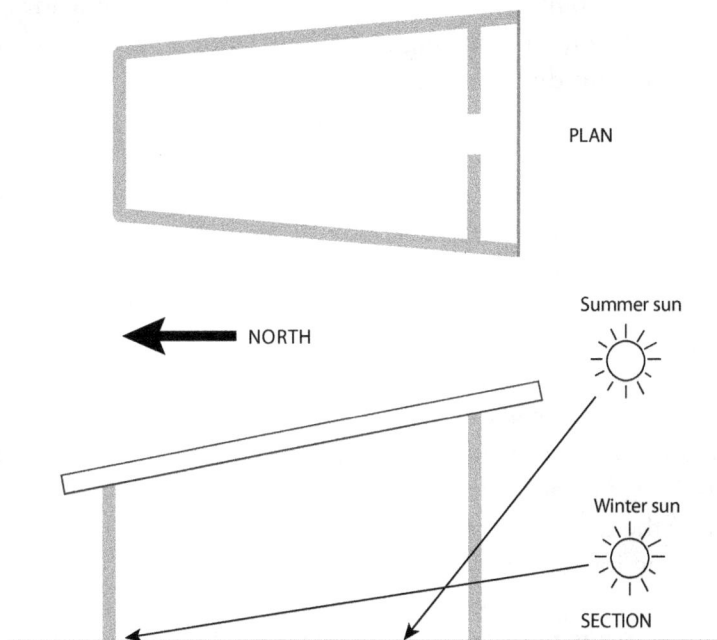

Fig. 13 Socrates "sun-tempered" house plan

Socrates' design sought to minimize excessive summer heat and winter cold by incorporating a smaller wall on the cold north side of the house than on the sunny south side, and putting deep eaves on the latter, which allow the winter sun to enter but blocked the undesirable heat of high summer, as shown in the figure above. Not without interest is the fact that this innovation came at a time when there was a great shortage in Greece of wood fuel for heating. Socrates, however, knew nothing of insulation technology and hence his trapezoid design (with monopitch roof and side walls sloping from south to north) is of decreasing architectural relevance as knowledge of insulation techniques and materials increases.

There was a ferment of architectural interest in the solar house in America in the 1950's and 1960's and many innovative designs. What all have in common is extensive use of glass, which enables solar energy to be collected and stored, sometimes in the internal structure of the building, sometimes by walls included only for this purpose, sometimes by putting rocks in the attic or cellar, normally using fans and ducting to draw surplus heat from them. Other countries came later to the challenge of the solar house, notably perhaps the Fraunhofer Institute for Solar Energy Systems in Freiburg, Germany. One commonly found feature is the incorporation of a massive concrete wall, sometimes but not always structural, referred to generically as a Trombe wall, after its inventor, the French pioneer Félix Trombe, mentioned above. It first appeared in 1962, though it had actually been patented in 1881 by the American Edward S, Morse and forgotten about. This type of solar house consists essentially of two layers of glass and and manual or thermostatically controlled louvres on the receptor walls to control the emission of heat. To this limited extent it could therefore be called an active system. A survey of the extensive literature on solar houses during this period is rewarding for the engineer-architect today, not least to avoid the mistakes that were made. The stock method of maximizing solar heat by increasing the amount of glazing drew the ire of Frank Lloyd Wright and others against what he called "boxes blazing in the sun." His solution, in part, was to go back to the Socrates house and extend the roof overhang on the sunny side, but it takes little imagination to see the inherent contradiction of deliberately reducing solar gain. The buzz of excitement in the solar heated house at this period is evidenced in the founding in 1955 of the World Symposium on Applied Solar Energy and a host of solar-based designs, from such as Fred Keck, his cooperator, the builder-developer Howard Sloan and Maria Telkes, around that time.

Subsequent developments reveal two factors of note in the context of today's global energy problem. Firstly, interest was driven by economics and fell away as the economic case for the solar house weakened. In his keynote address to the inaugural meeting of World Symposium on Applied Solar Energy the chairman emphasized that "social-conscious-wise" [his wording] there has been no demand for this kind of architectural innovation because it offered no competition to conventional sources of energy but, he went on to say, "these conditions are changing ....We realize, as never before, that our fossil fuels – coal, oil and gas – will not last forever."[5] At the time there was also the promise of electricity "too cheap to meter" from nuclear fuel generation, until the Three Mile Island disaster in 1979 put an end to that optimism. Interest which had been generated by the tripling of oil prices in 1973 by OPEC, died away as the world adjusted to higher energy prices. The second factor to emerge from these experimental designs, is the largely unacknowledged problem of energy storage, for which the E-plus house proposes a radically new solution.

The solar house of that period was, however, something of an architectural contradiction, in that the overriding criterion of garnering and (to a very limited extent) storing solar heat sat uncomfortably with the organization of space, which is usually regarded as the architect's main function. This presented an architectural Catch 22, summed up by Reyner Banham, "Narrowly preoccupied with innovations in the arts of structure, [architects] seem never to have observed that ....free-flowing interior spaces and open plans, as well as the visual interpenetration of indoor and outer space by way of vast areas of glass, all presuppose considerable expense of thermal power and/or air control, at the very least."[6]

A full analysis of developments during these years would call for several books, and would justify a university course, but one overarching factor is now becoming visible in the slow and patchy response of the building sector to the innovations in design and technology that were appearing then. In this regard, a notable, and successful, experiment in the UK calls for mention. In 1960 the far-sighted architect Emslie Morgan drew up plans for a school heated entirely by solar energy. Not only was it the first building in the world to be heated entirely by the sun, it was also (at 53.4°N) the most northerly and was conceived as a simple, self-regulating system. St George's school in Wallasey, which incorporated 1,000 square metres of glass and one windowless side, took Trombe's principles to the maximum by extensive ducting of warm air to balance the different heat requirements of the south and north side and incorporated insulation by plastic foam of a thickness regarded at the time as quite excessive. It worked faultlessly for five years, never having to call upon the auxiliary heating system even in winter, but Health and Safety Regulations decided (rather late in the day) that the ducting constituted a fire hazard, and modifications which were then carried out completely destroyed its critical function. Thereafter, there was over-heating in summer and the fallback oil-fired central heating system, which had previously never been used, was called upon in winter at an extravagant cost. Eventually the school moved to another site, and Morgan's masterpiece became a dysfunctional Grade II listed building. It is of particular interest that he was not a recognized *avant garde* architectural celebrity but only an assistant borough architect. He was unique, however, in having a passion before its time for utilizing solar energy. Despite the success of his school, which was all the more surprising in being a prototype, most of his detailed plans and all his complex calculations died with him in 1964.

Morgan probably took the principle of the passive solar building as far as it could go at the time. Given the scope of his imagination, his school might even qualify as active, rather than passive. However, inventions that have appeared since then have enabled a true active house to be conceived. In the past forty years they have come thick and fast, and the E-plus house either incorporates them or takes a long hard look at their potential for development and possible use either in the K-gen system or the house itself. The most obvious of these ground-breaking inventions are without doubt the photovoltaic cell, the solar thermal vacuum tube, greatly improved batteries, flywheel technology, the maglev bearing and the LED light bulb all of which have been assessed for their value in the K-gen system. Insofar as the E-plus depends for its function on the K-gen system, in which these and other inventions work are deliberately brought together, it could accurately be called in German the *Aktivhaus*.

## 10.4 The House as Energy Collector

At the heart of the new architecture are energy collection and conservation. The architect needs to get some firm ideas from the engineers before starting on design and, more than this, to internalise them and make the ideas a part of his or her working vocabulary. The following suggestions are necessarily incomplete but will, hopefully, provide sufficient information to indicate in outline the nature of the architect's future brief.

## 10.41 Solar Energy Collection

The "solar houses" mentioned above used two quite different principles in utilizing solar energy, neither of which appears to be capable of development, and certainly not of mass production. The revolving house design was based on following the sun, the others on having a large south-facing window enabling an interior concrete wall to store the sun's heat and release it later as required through louvres. There are literally dozens of such designs in the literature of domestic architecture, and while these fairly low-tech options are not to be ignored, more sophisticated engineering technologies must now be explored.

The two main ways of collecting solar energy already mentioned are by using photovoltaic panels to convert sunlight directly into electricity and solar thermal panels (tube or flat plate) to produce hot water. These are complementary technologies and not mutually exclusive. The use of roof-mounted parabolic reflecting troughs has by no means been ruled out, but introduces factors too complex to bring in here. Maximisation of solar energy collection could be achieved by using the roof area only for solar thermal panels, since they are six to eight times more efficient in collecting solar energy than PV panels, albeit in the form of hot water, and are relatively insensitive to the direction of the sun. A square metre of solar thermal panels could input the heat equivalent of about four or five kilowatthours on a decently sunny day. If these are installed on a reasonably south facing roof, freestanding PV panels can be put in the garden with optimum solar orientation or, preferably, with heliostat tracking mechanisms. The engineering challenge for solar thermal panels is finding a way to convert low grade heat energy from hot water into electricity and the method proposed in the previous chapter has been to make use of the organic Rankine engine. This does not exhaust the possible solutions to what is now coming to be seen as a well-defined engineering problem. While the sun is shining, a concentrating lens can be brought into use, which would raise the output water well above boiling, but the fact that it is already just below boiling point, or even above, means that it would sometimes be cost effective to use batteries or even mains electricity to create the expansionary function of steam or steam equivalent in a working fluid.

## 10.42  Wind Energy Collection

Using the roof of the house as a collecting area for a horizontally mounted wind turbine on the peak has already been mentioned as an engineering task, but the architect still has a role to play in designing roofs with the deliberate intention of making them more efficient collectors. Too flat a roof would present too small a collecting face, too steep would result in the wind impacting the turbine blades at an inefficient angle. Some simple mathematics would find the ideal balance, and could be tested out easily in a wind tunnel. As earlier mentioned, extending the roof downward to create a deep eaves would also in principle add to the collecting area. A channelling effect could also be created by building raised side walls, perhaps a metre high, which would minimize spillover over at the edge of the roof and add significantly to the amount of wind impacting the turbine blade. How far such a structural extension would detract from the aesthetics is a latent problem.

A quite different solution has been proposed for tall buildings, by leaving apertures in the upper part of the wall facing the prevailing wind, and installing the turbine itself within the roof space. [7] This may have merit in the case of a tall apartment block, offering the obvious advantage of increased wind collection at height.

## 10.5  Managing Light

The E-plus is envisaged as a house of light, seeking maximum sunlight collection.[8], bringing it into the house, distributing it, controlling both natural and artificial light and aiming for a balance between economic cost and all these functions. This approach is in keeping with the overall aim of an actively managed house. There is a great deal of literature on architectural lighting, but mostly relevant to large buildings and public spaces, with little or nothing about renewable energy sourcing nor psychotherapeutic factors, which are very much the concern of the E-plus house. Control of light can be approached in several ways, but the ever-present problem is how to maximize solar light without creating undesirable solar heat. Until a century or two ago the problem of natural lighting was defined by the cost and technology of glass making. Sigfried Giedion notes how in the Roman and late Christian eras "window openings became larger and larger," a trend which culminated with "flooding the interior with light [that] was characteristic of eighteenth century architecture."[9] He cites the town hall of Nancy (1755) as a prime example, which could be seen almost as an early form of curtain walling, both encountering the same problems of structural support and an inherent tendency for the interior to become a solar oven in very bright sunshine. An E-plus design would seek to find a balance, perhaps as simple as thermostatically controlled blinds. It would also exploit the concept of borrowed light, first emphasized in Sir John Soane's designs towards the end of the eighteenth century, and seek to complement it with strategically positioned reflective surfaces and modern technology, in the form of the sun pipe, or sun tube, which can direct sunlight with accuracy even into cellars.

The E-plus concept embodies five particular principles to increase internal light, natural and artificial. None of these is in itself new, but all are capable of significant development. The first is systematic use of the atrium principle; the second is deliberate use of reflective surfaces, as just noted; the third is wide angled window reveals, which offer both cost and insulation benefits as well as allowing more light into the room; the fourth is imaginative development of new cheap forms of artificial lighting, specifically the LED bulb, and the fifth is structural use of recently invented translucent materials.

## 10.51  Applying the Atrium Principle

Use of the atrium is currently almost entirely confined to commercial buildings, but if the principle were to be applied domestically it would effectively give the house an internal as well as external source of light. The domestic atrium, albeit unglazed, was actually a defining feature of the Roman *domus*, but has been strangely neglected in modern domestic architecture. It offers so much advantage in terms of light and all-round "liveableness", however, that it surely justifies radical reconsideration.

Three comments are worth making in the context of the energy generating house. Firstly, a choice of the atrium layout necessarily restricts or obviates other desirable options. It would, for instance, seriously inhibit the use of a roof peak wind turbine, but that may not be an option anyway in urban or sheltered areas. Secondly, it would be impossible to implement this sort of arrangement in the great majority of existing houses, even with a most radical refit. Thirdly, insofar as the domestic atrium is a principle that can be developed in several ways, it opens up the possibility of creating an E-plus portfolio made up of several possible arrangements of features, which could act as design templates. A reference work of this kind would simplify the work of domestic architects considerably. It should be clear by now that there is an enormous amount of uncollated information in the field of renewable energy, and it would be counter-productive for each individual architect to work out from scratch the sort of general or specific conclusions contained in the present work. Without such a database of architectural experience for easy reference there would always be a danger of reinventing the wheel of domestic microgeneration.

The following diagram represents one of several possible configurations, all incorporating mirrors, or other reflective surfaces.

**Path of Sun**

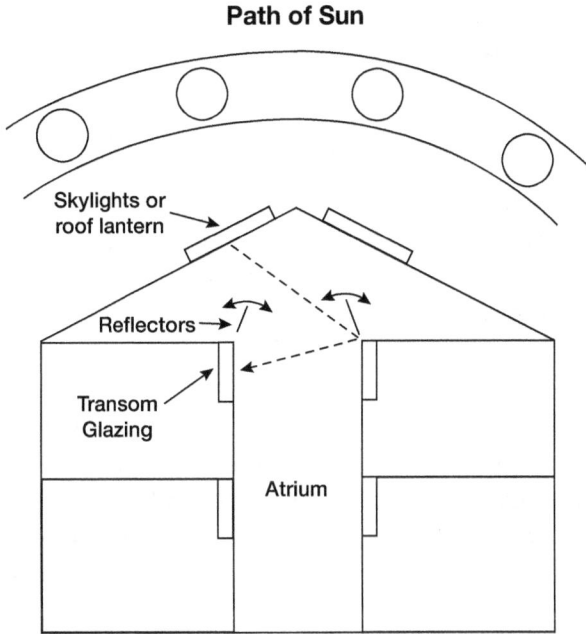

Fig. 14 Simple atrium schematic

## 10.52 Amplifying Through Reflection

Mention of reflectors as a means of transmitting external light to the building's interior, raises a more general issue which it is probably fair to say has been almost entirely ignored in architectural theory and practice. Externally, reflectors have been used as a way to boost output from photovoltaic panels, but the literature on this is rather scanty and fragmented and, *prima facie*, there would seem to be more gain by using heliostat mechanisms, rather than static mirrors. There have been experiments in America with different types of sun-tracking mirrors, usually directing the sunlight through windows and it could be worthwhile to re-examine the practicality of this apparently marginal idea, for it would be in practice make any wall south-facing. Painting the internal side of a garden wall white will throw back to the house a significant amount of light, which would be particularly welcome on a winter's day, or even constructing a white fence near the house for this specific purpose. The inventor-architect Steve Baer has used light coloured aprons as reflectors in front of the house to perform this function, an idea so simple that it seems to have gone largely unnoticed. Amplifying sunlight through reflection may seem to offer largely psychological benefits, rather than extra heating, but would be well worth having for that reason. However, once the broad concept of using reflected light is considered in depth, it raises issues about the size and placement of windows and these could be a significant factor in some architectural plans.

Interior reflected light, as touched upon in Figure 14, offers considerable scope for architectural imagination, especially when considered in combination with advantages in translucent technology. How light or dark the interior of a house should be is, of course, a very personal taste, but for those who would want to maximize interior light there are many imaginative ways in which it can be amplified, bent and directed through internal windows and surfaces which are in varying degrees reflective, rather than absorptive.

## 10.53 Augmenting Window Apertures

Angled window reveals were a not uncommon feature of large houses a couple of centuries ago, but rarely used in cheaper housing. To the obvious benefit of increasing the actual amount of light allowed into the room may be added the use of reflective material to surface the reveal. There are clear advantages to be gained when set against the initial expense of the slightly more complicated brickwork that is required. The actual window size can be reduced by 10-15 per cent to admit the same amount of light, and thus roughly *pro rata* the cost of the window, a significant saving. The improvement in light collection and distribution can be seen in the following diagram.

Direct light      Direct light

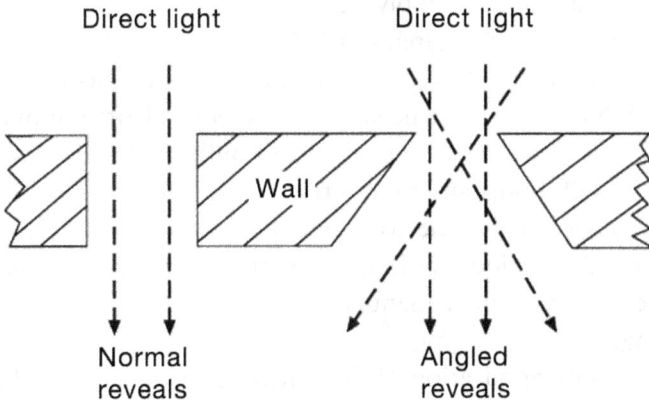

Normal
reveals      Angled
reveals

Fig. 15 Angled window reveals (reflective surface effect not shown)

At the same time, and particularly on thick walls, insulating the reveal improves the overall insulating of the house, given that even triple glazing is not a good insulator relative to, say, aerogel-backed panels.

### 10.54 Using LED Lighting

As regards artificial lighting, the economies to be gained from LED (light emitting diode) bulbs open up two architectural possibilities not previously considered because of the prohibitive cost of electricity using conventional tungsten or fluorescent bulbs. Research into LED is one of the most active frontiers of physical chemistry, and several cities throughout the world are already using LED lamps for street lighting, with electricity saving of up to 80 per cent over conventional argon or mercury vapour lamps. Similar and greater savings are claimed from domestic LED bulbs, and Professor Colin Humphrey at Cambridge University, who has pioneered development of gallium nitride semiconductors, has calculated that replacement of conventional lighting by LED's would reduce the national cost of electricity for lighting by three quarters and eliminate the need for eight big power stations. If these claims can be justified, it would be a very significant step towards reducing carbon dioxide emission. LED lighting comes with certain disadvantages, not least that most bulbs cannot simply be screwed into existing light systems, though this is changing, but it has one overriding advantage in cost-effectiveness.[10] Typically an LED bulb uses considerably less than 5% of the power needed for the equivalent illumination from an incandescent bulb, and the LED bulb itself will last in the region of 30,000 hours and more - that is to say, ten or fifteen years of normal usage. When one considers the cost of replacing normal bulbs over such a period, not to mention the wastage of material and embedded energy in their manufacture, the case for LED becomes compelling.

Whole house application of the principles involved would need to be carefully planned, since LED light is monochromatic and unidirectional (somewhat laser-like), ideal for desk lights but not for the sort of diffused light required for normal room lighting. These disadvantages can, however, be overcome in various ways, notably by the use of colour filters and diffusers, and this, in combination with the overall cheapness of the bulbs, opens up the possibility of having all day quasi-natural lighting during the winter months, using panels or arrays which emit light in the natural spectrum. LED technology is moving ahead rapidly, "Wi-Fi enabled" and switched on or off from a smart phone.

The importance of sunlight for physical and mental well being is rapidly gaining recognition, as so-called seasonal affective disorder (SAD), due to lack of winter sunlight, has been found to have a depressive effect on several of the body's systems, not least on the release of serotonin in the brain. SAD is already treated medically by using wall-mounted and desktop panels with filters to produce "warm" natural sunlight conditions from artificial lighting. Since LED lighting is so cheap, it would cost very little to gain these psychological and physical benefits from a dedicated, health-boosting lighting system to every house. The capital cost, including automatic switching, might be in the region of £2,000 to £4,000 and running costs for whole house, permanent lighting during the darkest four months of the year probably no more than £50. This opens up a whole new area for architectural innovation, since colour balance in general has wide significance in bodily response. We wake up more easily to blue light and relax more with yellow or orange saturated light. These are matters which impinge on the E-plus concept and design, and at this point the architect would need to bring in specialist assistance. Eventually state of the art becomes normal practice, and one can anticipate a time in the not too distant future when every architect would graduate with training in the new lighting technology and its effects.

### 10.55  Exploiting Translucent Aerogels

The insulating properties of aerogel have been introduced earlier, but it has other properties that make it of particular value when it comes to introducing natural light into buildings. It not only insulates acoustically, as well as thermally, but diffuses the light, thus eliminating glare and excess heat. It can be manufactured in large panels as the "filling" in a structural sandwich between glass or polycarbonate, with metal or reinforced fibre frames to give integral strength. Thus it marks an architectural breakthrough in several respects. Translucent walls 5-8m in height and undergirded roof spans of 20m are not uncommon. It is already being used to dramatic effect in public buildings, from churches to airports and factories and, given its extremely high U-value, it is natural to consider its use in domestic architecture.

Given the unusual properties of aerogel, the opportunity for its architectural use is limited only by imagination, but must begin with an in-depth and expert study of the technical literature of the few companies currently active in research and manufacture. Probably the most prominent in a rapidly expanding market are Kalwall®, Nanogel®, Thermablok® and and Cabot Aerogel, all from international companies with specialist UK suppliers. Little more than a mention can be given in this short section of the book, to the enhancement of natural lighting and insulation now available through aerogel-based materials. While the benefits have been conclusively proved in large buildings, and a few large houses designed for individual clients, to create what is sometimes referred to as "museum quality" light, there is reason to think that aerogels have a role to play in mass-producible housing design or in retrofitting existing housing stock.

The following four suggestions may be worth pursuing in the context of the E-plus house, partly to improve the quality of the environment, partly to improve thermal efficiency. Firstly, since aerogel material is now widely used for primary roofing, as well as skylighting (in America, at least), it is clear that any house design that makes use of the atrium principle can potentially benefit from it. Secondly, the internal diffusion of light which has been indicated in the diagram above only by transom glazing could be extended through whole wall aerogel panelling, which has significant load-bearing strength. Thirdly, since double and even triple glazing have limited insulating effect, there may be situations where this could be improved, albeit with the loss of some direct light, by combining it with aerogel panels in various ways. Fourthly, but probably not finally, its structural strength enables it to be used for both exterior and interior walls, offering potential for a clerestory or well of light effect. While this may at first seem inappropriate in a small house, it could be a simple way to eliminate long dark corridors.

## 10.6   Energy Conservation

The range of options in whole house insulation has already been in-dicated among the engineering tasks, but two particular architectural aspects call for mention. The first concerns cavity wall construction, which has been a Building Regulations requirement in most brick and block construction since 1991. Though obviously more expen-sive than single skin brick construction, cavity walling has been for well over a century the normal solution to penetrating damp of brickwork and reduction of heat loss through the convection ef-fect of moving air. It is, however, a low-efficiency method of house insulation in itself, and even when augmented by cavity insulation is far from ideal. With the invention of insulating and water-resistant polymer render, the self-insulated panel (SIP) and aerogel insula-tion, a far more efficient and cheaper option is now available in the form of single skin construction relying mostly on internal insula-tion. There are, indeed, overwhelming reasons for making internally and externally insulated single skin construction a future norm. The E-plus ideal would be to use aerated concrete block, aircrete, or SIPs instead of standard brick. Aircrete is a particular form of aerated concrete block made with up to 80% of pulverised waste from coal burning power stations. Both forms have structural and compressive strength and good sound and thermal insulation properties. A combination of aircrete with external insulating render and internal aerogel-backed panels would be near ideal as heat insulation and even better if the latter were to include a reflective foil layer. Viewed thus, the advantage of having internal insulation, rather than cavity insulation, become very obvious. Enthusiasm must be tempered, as usual, when considering the compromises that must be made in applying the principle to retrofitted systems, for many existing houses have very small rooms, and reducing that space by adding, say, 50 mm of insulation internally may be unacceptable. That said, if radical refurbishment is decided upon, hacking existing plaster down to bare brick and using 25 mm of aerogel lining or panelling extends the possibilities. This is the sort of expert advice that the architect or surveyor of the future should be able to give, and to cost out, as a routine service.

The principle at issue has further implications, for in a conventional house, the internal walls become part of the space to be heated, and not only take up a significant amount of energy but increase the time for the living space to reach the right temperature when heating is switched on and then wastes all this unnecessary heat in cooling, when the heating is switched off. By very rough estimate, 5 kWh of energy would be wasted in raising two or three tons of internal brick wall through fifteen degrees on a winter's day. In a well insulated house the temperature difference would, of course, be much less, but the broad principle remains, that there is no point in using energy to heat interior walls. Thus some thought should be given to the best material to be used for this purpose, bearing in mind that a degree of sound proofing will be needed in some areas of the house. Aircrete block would be a very strong candidate, particularly as it is a good acoustic insulator, but conventional studded and drylined walls may be a more economical solution.

Another aspect of the architect's brief would be to conserve energy through integrating into the design a waste heat recovery system. To flush away with the sewage almost all the energy used to heat domestic hot water for showers, laundry and dishwashing is eco-blindness of a high order. Utilizing it, however, is a specialized task, not least because a static heat exchanger always runs the danger of harbouring the bacteria of Legionnaire's disease, and there are already legal planning requirements to obviate this. The most effective waste heat recovery system must be planned in from the start, for not much can normally be added to a plumbing and ventilation system already in place. That said, there is at least one off-the-shelf system for utilizing the heat of waste water from the bathroom. Without doubt there is already architectural source material where this kind of information can be found, and the architect's brief in bringing the E-plus concept to realisation will be to collate and assess it.

The amount of energy lost by the average house in getting hot water to the taps can be very considerable and warrants particular attention. It is, of course, taken for granted that one should minimize the pipe runs for hot water and lag the pipes but when the actual figures are looked at more closely, the importance of this factor stands out more clearly. Precise figures can be obtained from AECB Water Standards,[11] but the following brief extrapolations from their data will be sufficient to indicate what is at issue. In the first place choice must be made between plastic and copper piping, and while plastic is rapidly replacing copper, for reasons of economy, flexibility and convenience, small bore plastic piping has a considerable disadvantage in reducing the flow through friction and requiring thicker walls. With this as background, the following specimen figures show the amount of water contained in a one metre length of pipes of different external diameters.

| Plastic: | 10mm | - | 0.03 litres |
|----------|------|---|-------------|
|          | 22mm | - | 0.24        |
| Copper:  | 10mm | - | 0.6         |
|          | 22mm | - | 0.32        |

From the perspective of energy conservation the significance of these figures can be appreciated by considering the last item, which would be about the dimension of pipe required to take hot water from a tank or boiler to feed a shower, allowing for some take-off along the way to other hot water taps. If the bath is located fairly close to the tank – say, in the room above but at the opposite end – it is easy to calculate the amount of hot water that will remain in the pipe when the shower is turned off and thus the amount of energy that will be wasted when it cools. Taking a medium size room to be 4.5m long and the drop in height to be 2.5m, each time the shower is used it will leave (4.5 + 2.5)m x 0.32 litres of water to cool in the pipe, that is, 3.6 litres. If the bath or shower were further away from the hot water tank, the figure could easily increase to 10 litres. The cost of heating this amount of unused water can be considerable, and it happens every time dishes are washed or someone runs the hot water for a shave. Exact figures are, of course, not possible, but one can expect the average family to waste every week the equivalent of the energy required to boil twenty or thirty electric kettles.

It is not something that figures much, if at all, in the domestic architect's brief, but hotel designers are certainly aware of it, for they must ensure that every room has immediate hot water, while keeping costs down. Those guests in older hotels who have waited half a minute for water to run hot will be aware of the problem only as an inconvenience. Modern design solves it by, in effect, extending the boiler throughout the building as a large bore, very well insulated pipe incorporating pumped return to the boiler, with small bore spurs running to each room. This is not appropriate for the average house, of course, but reflecting on its purposes and drawbacks may serve as a starting point to think more creatively about this form of energy wastage as a factor in house design. Two obvious principles emerging are optimum location of the hot tank and superinsulation of all hot water pipes.

Ironically, pursuing the E-plus ideal leads backwards towards some common features of Victorian house design, particularly when the problem of energy storage is tackled. Where formerly a cellar was routinely incorporated to store coal, a tanked and superinsulated basement could be used for heat energy storage, or for batteries or flywheels once it has been decided how the heat or electricity is to be stored and released. The risk associated with flywheel "explosion" has to be considered and installing one safely in a basement will call for some expertise from a structural engineer. However, it is probable, as already noted, that the optimum scale for flywheel energy storage may be of a size to serve several houses, and consideration of this will have an impact on the configuration of a newly designed Grid.

A second feature to be reconsidered is whether the semi-basement, which is standard in North America and other countries, is preferable to a cellar as such, that is to say, where about a quarter of the basement height is above ground level, thus providing natural lighting. This would result in something like a 3.25 storey town house design that would offer additional energy-saving, especially if the roof space were to be designed as living space. A semi-basement adds valuable space to a house, a consideration which becomes of increasing importance when the nation's demand for building land is becoming critical. If the windows take up the top metre of its height, a semi-basement can be used as extra living space as well as for utilities and storage. The old-fashioned cellar fell out of favour fifty years ago, as oil and gas central heating became the norm and the need for coal storage space disappeared. Without a cellar a very significant saving in excavation cost could be avoided, often over 15% of total building cost, but the time has surely come when this initial saving needs to be weighed against long term gain.

For lack of under-roof insulation, roof space is an architectural feature that has never been properly developed, and in most houses the attic is left as unheated storage space. Over fifty years the trend in domestic design has been to minimize construction cost by using hipped roofs, which unfortunately minimizes under-roof space. Furthermore, prefabricated trusses have replaced the traditional attic and mansard trusses, ostensibly saving the cost of heavy timbers for purlins and rafters, but filling most of the roof space with a criss-crossing bird's nest of small spars for half the cost of material. This is, however, a false economy when one considers the value that could be added by having extra living space. It is also worth mentioning that the extra height of a taller house, as against the standard "semi", would give substantial advantage in wind energy harvesting, since wind speed increases sharply at about ten metres above the ground. Another traditional feature that could well make a reappearance is the larder, built on the north or northeast side of the house, but now cooled by the waste cold air from the heat pump. Little of the food inside a refrigerator needs chilling below 5 degrees and this could easily be maintained in a larder, which would be a walk-in cool room, storing most of the medium- and long-dated food that is currently kept in an over-warm kitchen. This idea is, in fact, already being marketed by one company, which offers its heat pumps with the option of a ducted recirculating air system taking air to the installation room, which their literature states can then double as "a cool storage room for groceries".

## 10.7   New Approaches to Central Heating

The importance of the heat pump in maximizing energy production has already been emphasized and has significant implications for central heating, since it focuses attention on the problem of how best to utilize the (relatively) low temperature water which produced by the heat pump. Domestic heating has evolved from the open fire to central heating using a coal-fired boiler with heavy cast iron radiators then to the oil- or gas-fired boiler with mild steel flat radiators and within the last half century to finned steel radiators. All these operate with water at about 70°C, though old wide bore systems in public buildings used very hot water, often approaching 85°. With the advent of underfloor heating, the heat pump can be used as a central heating boiler, utilizing output water at much lower temperatures, typically 45°. As the previous chapter explained, heat pump efficiency increases as the difference in temperature between input and output water decreases; so the challenge to the design engineer is to raise the former and, where possible, find ways of using water at lower than traditional temperatures. Solutions to date have been to use ground source heat as input and underfloor heating as output, thus providing mild continuous heat with excellent space heating profile - that is to say, with minimum hot or cold spots in the room.

There are, however, serious drawbacks to this solution. The capital cost and upheaval of installing a ground source system have already been noted, but there are also disadvantages arising from the fact that the warming effect of underfloor heating falls off when the floor covering is not a hard smooth surface but a natural insulator like a carpet and also from the fact that it can be normally installed only on the ground floor. These disadvantages can be overcome by an air source, rather than ground source, heat pump, using prewarmed air from waste heat to raise the input temperature and by increasing the rate of heat transmission at the active face of the central heating system. It is easy to see in a general way that one minute of heating at 10° from a panel a metre square will transmit roughly the same amount of heat in the same time as a panel of half a square metre at a temperature of 20°. If this fact be linked with the principle that the coefficient of performance of a heat pump is an inverse function of the temperature differential between input and output water, a new perspective on space heating is gained. The efficiency of a heat pump used as a boiler, as in underfloor heating, can be effectively doubled if the input-output temperature difference can be reduced from, say, 7°-60° to 12°-45°. Following this logic, three specific recommendations can be made, all tested and all capable of further development:

- utilizing the waste heat of the house to raise the temperature of input air through conventional methods of jacketing, ducting and heat exchangers.

- increasing the rate of heat transmission of the heating surface of radiators by replacing steel with aluminium and copper, which are far better heat conductors

- increasing the rate of heat transmission area by increasing the active face, through dense finning or with metal skirting board heating instead of conventional wall mounted radiators.[12]

Whereas it has always been normal practice to design the house and treat the heating system as an add-on, the architect will now need to start with only a general idea of shape, size and layout in mind, and begin his or her task by specifying the energy management system in some detail before proceeding to what previously would have been considered architectural design proper. It should also be pointed out that as Passivhaus standards of insulation are approached, the need for central heating diminishes, so that a very low capacity system, driven by a small heat pump, would be adequate, and required only in the coldest weather. This back-to-front approach to design will be possible only with new-builds, and the greater challenge will come, as always, from refurbishing existing housing stock to give maximum energy efficiency.

## 10.8    The Retrofit Challenge

If the E-plus concept is to make the anticipated impact on reducing energy demand and providing base load electricity, an optimal balance must be found between the cost and efficiency of every retrofitted installation. Confronted with the small Victorian terrace house, of which there are probably six or seven million in the UK, the dreaming has to stop, for the cost of a radical refurbishment to maximize energy efficiency is likely to run to half the value of the house, and the gain could well be minimal. It is not easy to make general recommendations for a situation where every house has its own characteristics in almost every respect, and ideally calls for an individual report and cost-assessment from a specially trained surveyor with substantial experience in design, quantity surveying and strength of materials.

Surveys of this kind are already available for eco-refurbishment, based largely on Passivhaus principles, and Sustainable Energy Ireland offers a comprehensive sixty page document, "Retrofitted Passive Homes", available as a PDF, with useful data of all kinds. On examination, a general principle emerges that the more efficiency required, the closer the retrofit comes to total gutting of the original house, which is rarely an option, not only because of the cost but of the inconvenience.[13] There must always be a temptation to rebuild when retrofitting, rather than work around existing structures, but cost-effectiveness in energy usage calls for a more subtle approach. In some, perhaps even most, instances a total strip-out may well be the preferred choice, particularly if the house is rather decrepit to start with. In others a patchwork of partial solutions may be called for. Efficient insulation is the first criterion, as has been emphasized, and the architectural focus should be on internal insulation, but thought needs to be given to zone heating, for in practice few rooms in a house need to be kept at living room temperature. This is to some extent catered for by individual radiator thermostats, but a normal bathroom would, for instance, be best served by instant heat provided by a quartz or fan heater. Lateral thinking may be called for in many cases because lack of space in a small terraced house obviates many theoretical solutions. There may be something to be said, for instance, for old-fashioned solutions to problems like drying laundry not in a tumble dryer but by using a ceiling rack in conjunction with a dehumidifier, which generates heat as well as extracting moisture.

When all the difficulties are taken into account, however, there is still a body of expertise to be taken from new-build technology, as sketched in this chapter, and applied to retrofitting. Simply rethinking the principles of lighting and insulation from scratch will almost always throw up possibilities for enhancing the comfort and energy-efficiency of any house.

### 10.9   Harvesting Water

While the book deals with the global energy crisis now looming, a parallel crisis is now coming to public attention in the form of increasing water shortage across the planet and this has architectural and engineering implications which call at least for mention. There are, in fact, some striking parallels with the E-plus concept insofar as harvesting both energy and water can be approached either as a matter of simple add-ons or as systems with varying degrees of sophistication. Also, maximum efficiency can only be achieved when designing and building a house from scratch, whereas retrofitting an existing building inevitably calls for compromises.

The background to water shortages in the UK is a combination of four main factors. Population increase has raised demand, but in an uneven way, with the south east quadrant of England making exceptional demands. The infrastructure of water transmission and sewage disposal is in a state of increasing decay, most of it having been laid down anything from 80 to 150 years ago, resulting now in serious leakage, often 15% or even more of the water en route to the consumer. Less quantifiable are erratic rainfall patterns due to global warming, with growing risk of periodic drought conditions which had been masked when the population of the UK was smaller. Increasing demand for water has come with growing prosperity in the latter part of the 20th century, expansion of suburbia bringing with it gardens to be watered and cars to be washed. Overall, the average daily water consumption per person in the UK has risen from 100 to 150 litres in the past 25 years.

The answer to this growing water shortage lies, quite simply, in every house conserving water and collecting the rain in exactly the same way that the answer to the UK's energy shortage lies in every house conserving energy and collecting solar and wind energy. Domestic water collection has been commonplace for a long time, particularly in rural areas, by diverting the rain that falls on the roof into a water butt, which has traditionally been used to water the garden. Sometimes the water was removed with a scoop or bucket, but sometimes the butt had a tap at the base from which a watering can could be filled. The cost of this arrangement was negligible, but so too was the effect on the reducing the country's need for mains water. Modern rainwater harvesting systems (RWH) can be at several levels of complexity and expense, but all are variations of the simple downspout and butt arrangement.

The next level is a large tank with integrated pump to take the collected water into the house. However, to ensure a continuous supply of domestic water, a very large tank is required to hold several weeks of reserve, which is usually unsightly and even in a shaded position will collect algae and other growths in warm weather. A better solution is to bury the tank, where conditions allow, adding a pump to cope with overflow conditions, which the above ground tank can handle just by emptying excess into the public drain. Excavating and burying the tank has the great advantage of keeping the water cool and thus inhibiting growth, but a periodic flushing and disinfection is still necessary, as well as filters.

Assuming an underground tank, the next level of sophistication comes in deciding how the collected water is to be used in the house. The harvested water can be used as it arrives, after simple filtering, for flushing toilets and washing clothes, but additional uses call for some thinking about the degree of purity required. Apart from contamination that may occur in the tank, rainwater not only contains pollutants that it picks up from the atmosphere, but flows over a roof that is often liberally covered with bird droppings, especially under TV aerials. The seemingly simple problem of harvesting water thus soon runs into quite separate problems of filtration and purification. Specialist advice is called for in dealing with these, and there is no shortage of it on the Internet, where several competing RWH systems are on offer, with different levels of sophistication. All that need be said here is that the raw water can be purified to mains quality or higher by activated carbon filters and to the highest drinking water quality by a reverse osmosis system (costing about £250). Where harvested water is to be used within the house, rather than externally for the garden, etc., the Water Supply Regulations (1999) call for inspection and approval by the mains supplier. Where it is to be used only to water the garden and wash the car, no regulations apply and a gravity fed system, such as offered by Rainwater Hub, is adequate and has the advantage of being simple and virtually maintenance-free, with no pumps, and also modular, so that it can be expanded as experience dictates requirements. For the latter reason, there may well be unseen developmental potential in such an unsophisticated system, since an ordinary garden, or even a back yard, could easily provide above ground storage for 5-10,000 litres of water. Viewed thus, new possibilities become apparent, as, for instance, coupling this to a mini-treatment system, imitating the procedures of water companies. This would have a particular value in hard water areas.

The highest level of water conservation comes from reusing waste water wherever possible, which calls for a system within a system. So-called "grey water" from lightly polluted sources, such as wash basins, baths and showers can be used to flush toilets and irrigate the garden, saving a quarter to a third of household consumption. It does, however, require dedicated plumbing and additional pumps and, though theoretically ideal, is by no means always practical or cost-effective. At the same time, it is worth considering that with the obvious exception of sewage waste (the recyclability of which is a challenge for the future) all the water consumed by a household is capable in theory of being recycled. Since run-off water from farms, which is now heavily polluted with fertilizers and pesticides, can be made drinkable by the water companies and astronauts can survive for months in space on recycled urine, there is good reason to take a radically new look at the possibilities of recycling domestic water as well as harvesting rain.

As regards the economics of RWH, only the roughest figures can be provided, since the whole scene is changing rapidly and available systems are of different specification and price. With this proviso, an RWH system with underground tank but without recycling function would cost about £4,000 and could be expected to save 40-50% on consumption, with a consequent saving on water bills of £100-£200 where usage is metered. The actual saving is not *pro rata*, since a standing charge is always payable even with metered water and, of course, there is no saving when the water is unmetered. Running costs for maintenance and electricity to power the pump, or pumps, might be about £40 per year. If these figures are within reasonable limits, an RWH system would be a good investment but the advantage always lies with a new-build house where the system can be integrated at the design stage. The UK Rainwater Harvesting Association calculates a payback period of about 10-15 years, but this is certainly conservative when one considers the added value to the house, not to mention other benefits such as soft water.

The water crisis is far more acute in countries other than the UK and as public awareness grows, there will no doubt be root and branch thinking about its solution. Obvious new avenues to be explored are the benefits to be gained from regarding the whole land area of the house as a potential collecting area and thinking systemically about the different qualities of water that the householder required. The former leads towards making use not only of the roof of the house but of outbuildings, greenhouses and freestanding solar panel installations for harvesting rain and the latter to a system of three filters graded according to the amount and purity of the water required. This latter would call for quite complex plumbing, coupled with pumping and gravity feed arrangements. From such a perspective there would be a natural tie in with the E-plus concept in two obvious respects, namely, using freestanding solar panels as water collectors, as just mentioned, and home-produced "free" electricity to power the pumps.

One aspect of water harvesting that can only be mentioned here but calls for a book-length treatment because of its importance is the extraction of water from the atmosphere. The technology for this is well known but its potential goes largely unrecognized, and the World Health Organization and charities like Water Aid, which campaign to make clean water supply a priority in developing countries, do not appear to mention it in their literature, so far as the writer is aware. Air contains anything from 1-4% of water vapour, which can be "squeezed" out by cooling below the dew point, when it condenses as water, using technology that is a slightly more sophisticated version of that which is used in domestic dehumidifiers. The technology exists on different scales. At one end, the US army regularly employs mobile units about the size of a bus powered by 30 kilowatt diesel electric generators, and capable of providing up to 5,000 litres of pure water per day. At the other end are domestic size extractors running on mains electricity. It needs no imagination to see how one of the latter could be linked up to a renewable energy source – PV panel or small rooftop wind turbine – to give a steady supply of very pure water to a household. Unsurprisingly, in a country like the UK, which has a humid climate, ample rainfall and mains water there has been no demand for water extracted from the atmosphere, but in other regions less well endowed if the existing technology were to be married to a renewable energy source, there could be a genuine revolution in water management. Even in the UK, a small scale extractor would do away with the need for bottled water, of which millions of litres are purchased daily by people who find the local tap water of unacceptable quality. The number of plastic bottles required for this, many of which finish up in landfill and in the sea, is one very serious side effect that would justify research in this area.

## References

1. Walter Gropius, *The New Architecture and the Bauhaus* (1925). Trans. P. Morton Shand, Cambridge MA: MIT, 1965. p. 89 & p.99.

2. The quote is slightly adapted from Gropius's invited article in the *Architectural Record* of March 1956, in which he talks of "struggling with the first letters of this new alphabet" at the beginning of the Bauhaus movement, takes pride in the fact that it is now accepted "as a matter of course" in the architectural world, but regrets that it has "not taken root yet within our American population as a whole." Nor did it in Britain and elsewhere. Given the principle of "the better mousetrap", the reasons for this are are of social and architectural interest.

3. Figures and quotation from Anthony Denzer, *The Solar House: Pioneering Sustainable Design*. New York: Rizzoli International Publications, 2013. p. 222.

4. As so often, Banham's almost gonzo style covers radical insight. The original article, now fifty years old, which uses the witticism as a title, opens, "When your house contains such a complex of piping, flues, ducts, wires, light, inlets, outlets … hif-fi reverberators, antennae, conduits, freezers, heaters – when it contains so many services that the hardware could stand up by itself without any assistance from the house, why have a house to hold it up …. What is the house doing except concealing your mechanical pudenda from the stares of folks on the sidewalk?" "A Home Is Not A House", *Art in America Number Two*, April, 1965.

5. Anthony Denzer, *The Solar House: Pioneering Sustainable Design*. NY: Rizzoli International Publications, 2013. p.47.

6. Quoted in Nigel Whiteley, *Reyner Banham: Historian of the Immediate Future.* Cambridge MA: MIT Press, 2002, p.366.

7. Professor David Infield, Director of Renewable Energy Technology at Strathclyde University has done extensive CFD modelling of embedded ducted turbines in tall building structures. The obvious attraction of this is that wind speed increases with height by and the increase is in proportion to the square of the speed.

8. One should add "at UK latitudes", for in a Mediterranean region thought would have to be given to minimizing the heat of the sun in summer. This qualifier brings home an important fact that a universal E-plus house is not possible, but must be designed for particular climatic conditions. An E-plus house in northern Scotland, for example, will not be able to rely on solar energy, and hence the designer would naturally tend to emphasize wind collection. In instances where there is a lack of either solar or wind energy, the E-plus design would effectively reduce to the Passivhaus plus intelligent use of heat pumps.

9. Sigfried Giedion, *Architecture and the Phenomena of Transition.* Harvard UP, 1971. p. 5.

10. See Scott Daniels, "The LED Lighting Revolution - a Primer" on the *Provide Your Own* website.

11. The Association for Environment Conscious Building (AECB), established in 1989, is a not-for-profit "network of individuals and companies with a common aim of promoting sustainable building." It aims for "excellence in design and construction, rather than gimmicks and green accounting tricks" and offers a "low energy buildings database." (From their website www.aecb.net ).

12. There are several types of heated skirting board in use, with different properties and costs. The more conventional is made of steel with louvres to emit hot air. Discrete Heating, a UK company, markets and installs a patented system made of aluminium, using low temperature water, which gives both convected and radiant heat and a better space heating profile.

13. See "UK's First Passivhaus Retrofit" on the website of Green Tomato Energy. The report says, it "clearly costs a lot more than a standard house refurbishment [since] the house was stripped back to its bare bones." There came a point when one could "stand at ground level of this four storey house and see the sky." Such a ruthless refit suggests that it might be preferable to demolish the house completely, as the only thing remaining would seem to be a shell of brickwork, which could be replaced with custom-designed aircrete blockwork and insulation at modest extra cost, perhaps £15,000. The case is strengthened by the tax anomaly through which VAT at 20% is levied on improvements but not on new-builds.

# Part Five

# The Scale of Things

*Reaching and passing the peak of world oil production will be the most important happening in human history to date, affecting more people in more ways than any other event. It will happen, and during the lives of most people now living.*

Walter Youngquist (1999).[1]

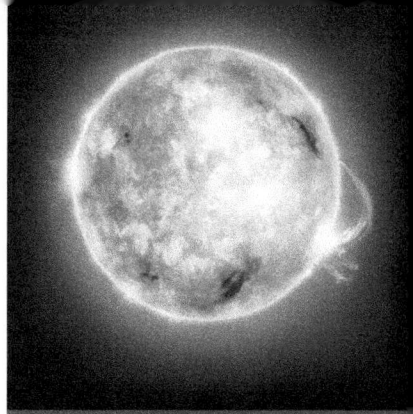

# Chapter 11

## An Engineering Revolution

*Technologies are not mere exterior aids but also interior transformations of consciousness.*

Walter Ong[2]

### 11.1 The Self-powered Engine

Engines for utilizing renewable energy have appeared in the past, two of which may be considered as a step change in the concept of engine, both named "sun engine" by their inventors. In 1878 Augustin Mouchot demonstrated at the Paris Universal Exhibition a printing press powered by solar energy. He focused the rays of a parabolic reflector onto a small steam engine, which generated enough energy to drive the press - so long as the sun was shining. He also set up industrial applications in Algeria, where the sun was stronger and more reliable. The second invention, by the American Frank Shuman, was designed on similar principles, but used solar troughs rather than parabolic reflectors and was on a scale large enough to pump over 20,000 litres of water a minute to irrigate crops in Egypt. That very significant amount of water was delivered without hitch from 1913-15, before the engine was dismantled, apparently to be used as scrap in a misguided attempt to help the war effort. Significantly, neither of these initiatives was further developed, since coal and oil had become cheaper and the sun engines stopped working when the sun stopped shining. The principles of the sun engine are, however, still valid, and Mouchot's book *Solar Heat and its Industrial Uses*, published in 1869, is still worth reading. His design for a solar engine is archetypal in its simplicity and is at the theoretical heart of massive arrays of solar generators now being used in desert conditions, the only significant difference being that they focus the sun's rays on Stirling engines rather than steam engines. If a machine is narrowly defined as a device for transmitting power, and an engine as a powered machine, Mouchot's invention can be seen as the start of a revolution, in that it is driven by renewable, available and cost-free "fuel". From this perspective the K-gen system calls for attention.

## 11.2 A New Concept of Engine
The K-gen system takes the concept of the self-powered or free-energy engine a critical step forward in four respects. It aims to

- harvest free energy from multiple sources
- store and deliver heat directly where needed
- convert heat and torque to electricity
- store and deliver electricity on demand

Figure 18 below brings into relief its main features and their interconnection. The box labelled "state change" refers largely to devices operating on the heat pump or Organic Rankine principle.

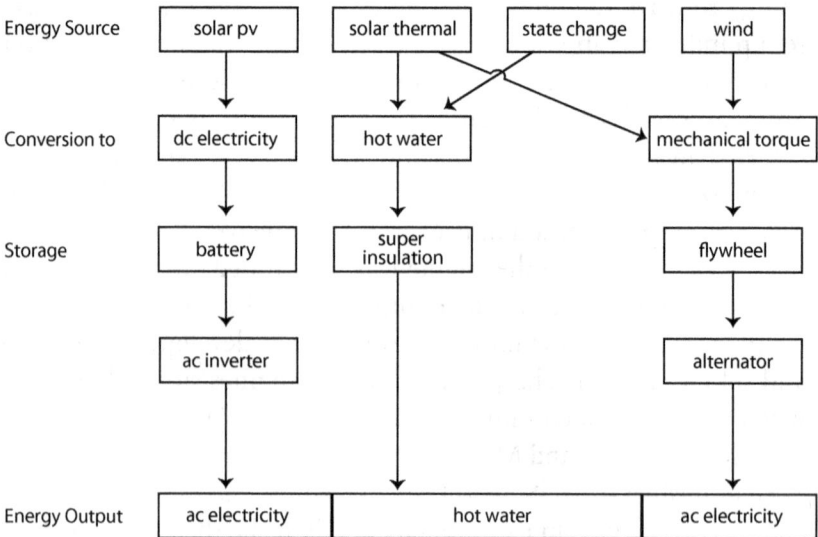

Fig. 16 Simplified schema of K-gen system, showing four functions.

The system itself is actually an engine, but of an innovative kind, since it brings together several forms of energy-generating devices which, acting in a coordinated way, will provide electricity when required and in the amount required. More significantly, it stores energy in several ways, as against external and internal combustion engines, which store it in the chemical bonds of the fuel. Viewed thus, the K-gen system can be seen as a new concept of engine and, just as the development of the petrol and diesel engines called for skills different from steam engineering, development of the K-gen system will call for different engineering skills, particularly perhaps for more than usual imagination. The reason for this is that an engineer, almost by definition, is a specialist in some particular field, and hence does not think naturally in terms of conceptual systems. There is, for instance, no reason why an expert in battery technology would be aware of advances in other forms of energy storage, let alone in wind turbines or solar panels. Hence he or she would have no interest in making the sort of connections, which are at the heart of systems engineering. The systems engineer needs to have both a wide horizon of knowledge and a critical degree of depth, plus a well developed curiosity and power of logic, intuitively asking "What if?" and "If this, then what?" The term "systems design engineer" is almost a self-contradiction insofar as he or she must be, as it were, a specialist in general engineering and at the same time both a practitioner and an imagineer. Imagineering is quite different from dreaming or brainstorming: it is, to use the original definition, "letting your imagination soar, and then engineering it down to earth."[3] It is perhaps worth noting another factor that works against development of the engine as system, namely that because a system by definition has several interacting components, it is more difficult to explain and justify to a grant-aiding body than a single-functioned device. A working prototype is needed, but this is precisely what the finance is needed for – a perfect Catch 22. It is hard to imagine the Wright brothers finding investment to develop their concept of a flying machine until they had been able to demonstrate a working version of it to a potential investor.

## 11.3 K-gen and the Anthropocene Era

Although the claims made for the K-gen system are exceptional, and remain to be proved, it marks only the beginning of a larger and mind-stretching revolution. Attention has been drawn in Chapter 1 to the way in which fossil fuel energy has literally become a geological force and changed the face of the earth, and enough has been said in this chapter to show that if a quasi-self-powered engine on the principles of the K-gen system can be constructed, it will have equally far-reaching, and balancing, consequences. The Nobel laureate Paul Crutzen, an atmospheric chemist, coined the term the "anthropocene era" some fifteen years ago to register the fact that changes in human behaviour, resulting almost entirely from the use of fossil fuels, have changed the geology of the planet and created new layers, or strata. This idea had been, so to speak, simmering for a long time, at least since Vladimir Vernadsky, a founding father of geochemistry, first proposed the ideas of biosphere and noosphere in 1926 and gave them the names which were later popularized by Teilhard de Chardin. Now the concept of an Anthropocene Era seems to be coming into focus with an international team of scientists, assembling as the Anthropocene Working Group, to give it scientific recognition. In the context of the global energy trap their aim is of particular interest, although it may at first seem a long way from the interests of engineering and architecture. The history of the earth is marked by clearly definable, though not precisely datable, stages when certain rocks were laid down, certain vegetation appeared and certain climatic conditions prevailed. On this scale we are at present approaching the end of the Holocene Era, which is among other things, an interglacial period. The start of the Holocene can be dated with some exactness to the type of vegetation appearing as the last ice age came to an end, which coincided with the end of the Old Stone Age stage of humans. Our tool-making species has evolved since then in varying ways, at first with no effect on the planet, but now with sufficient impact to call for a new classification. The point of what might at first seem a parlour game is that taxonomies are markers of significance and this is where earth science, energy science and human evolution come together. This is not the place to look further into the metascientific issues, but to illustrate the nature of the problem in dating the Anthropocene, it might be asked whether or not the urbanization of our species is now a geological fact, given, for instance, that greater Tokyo now has a population of 38,000,000.

The Anthropocene Working Group, which assembled in Berlin in 2014, are working to decide whether all the relevant facts now combine to justify official adoption of the term Anthropocene (and recognition of its significance) by the International Commission on Stratigraphy (akin to the SU, the International System of Units). In a word, would such a neologism deepen our understanding of science and the human condition? Clearly the present book would favour such a decision, but it would differ from the working group in dating the end of the Holocene and the beginning of the Anthropocene. The working group, so far, favours a very precise date, July 15, 1945, which marks the detonation of the atomic bomb in the New Mexico desert, but if the argument of the book is followed through, it would be more logical to go back to Thomas Savery's patent of 1698 for an invention "occasioning motion to all sorts of mill work by the impellent force of fire", or perhaps to Newcomen's patent in 1705 for his much superior "fire engine", or perhaps even brought forward to 1763, when Watts tweaked the basic concept with the condensing boiler, thus making the steam engine compact and powerful enough to used for transport as well as for pumping out mines. The development of the external combustion engine is a paradigm case of the incremental process that begins with no more than an idea, a hunch, a "what if?", and finishes with an affordable device that changes society and, as the quotation from Walter Ong at the head of this section asserts, transforms human consciousness in a fundamental way. It is unusual, and a cause for some national pride, that all three inventors named here were British, for a seminal new idea in engineering or related fields usually activates cooperation and competition across the civilized world. While a case for dating the the start of the Anthropocene Era to 1945 can without doubt be made, a stronger case can be made for the fossil fuel revolution which came to its peak with the invention of the petrol engine.

If the K-gen system, as outlined in the book, lives up to its promise and proves to be a generic concept for an engine without fuel, powered by renewable energy, which it then transforms into electricity and stores and delivers on demand, it would surely be the outstanding candidate as a marker for the beginning of a new and positive phase of the Anthropocene. It is, however, hardly likely to be so far developed when the Working Party presents its recommendations to the International Committee on Stratigraphy, planned for 2016.[4] Until then opinions are almost universally pessimistic, with E. O. Wilson's forthcoming book *The End of the Anthropocene* anticipating its end fairly soon, when the biosphere, "an extremely complex and razor-thin system [will] decay and unravel ... and the whole thing will collapse – and we collapse with it."[5] That is a succinct summary of the "global energy trap", but if the argument of the present book's is accepted, new vision in the philosophy and practice of engineering can in the longer term provide escape.

Thinking in terms of basic electrical supply from renewables will call for, and call forth, an epochal shift in engineering consciousness, as did the invention of the airplane, which depended upon and inspired new thinking in many fields, from fluid dynamics to petrol engine design. The K-gen system is, in fact, comparable in several ways to the first airplane. The basic concept, the *idée force*, of a machine which lifted itself by forward motion, rather than flapping like a bird, was inspired by a single principle (Bernoulli's theorem) and was at first counter-intuitive. In a similar way, the K-gen concept grows out of a single principle, called here the Return to Equilibrium, or RTE, which has already been mentioned and calls now for more attention.

### 11.4 The RTE Principle and its Application

The RTE principle is not new and is, in fact, is implicit in Carnot's theory, which initiated the science of thermodynamics. The task he faced was to create a quantified theory to explain and develop the recently invented steam engine, and this he based on two observations which he did not relate. The first was that the quantity of work done by the machine had a limit as the coal or wood was reduced to ash. From this insight Clausius was to develop the concept of entropy, which underlies the second law of thermodynamics and which asserts that spontaneous change in a closed system is always from order to disorder. Oversimply, the coal that powers the engine is energy in a concentrated and chemically ordered state, while the ash is chemically disordered. At the same time, Carnot noted another quite different limiting factor, namely, that every thermal engine depends on having a hot source and cold sink. The two factors have never been fully reconciled in heat theory, but the latter is the basis for the principle of RTE. Its significance lies in the fact that a hot and cold system in contact will spontaneously seek an equilibrium temperature and it is this seeking (called *streben* - literally, striving - by Clausius) which is the source, and perhaps even the reality, of energy. A hot and cold system in connection will create thermal pressure, and at extremely low or high temperatures will create extreme pressure. RTE explains, as the law of entropy cannot explain, why a Stirling engine can run equally well on a hot or cold source, as it seeks equilibrium with the ambient temperature. A model Stirling engine can run with ice as a "fuel" and will put out much more power if run on liquid oxygen or frozen carbon dioxide.

The importance of RTE in the present context of harnessing renewable energy is that its prompts the engineer to identify situations where there is an imbalance to be exploited (as, for instance, a water wheel exploits a gravitational difference) or to deliberately create such an imbalance. The available options are limited only by imagination, all depending on "extracting" energy from constantly changing temperature and air pressure levels throughout the day. The principle has been used for a long time in timekeeping technology. The Atmos clock, an outstanding example of Swiss craftsmanship, is driven by the power generated from a small bellows device which expands and contracts sufficiently with only one degree of temperature change per day to keep the clock running indefinitely. In one sense, however, the Atmos clock is a reinvention of the wheel, since a similar but larger clock using both air pressure and temperature change was invented by Arthur Beverley in 1864 and, without ever having been manually wound, is still running at the University of Otago in New Zealand. Even that was not a first, for the Victoria and Albert Museum in London has a clock (deactivated some years ago) made on the same principles by James Cox and John Merlin as long ago as the 1760's. The heart of the Beverley clock is a one cubic foot box which expands and contracts sufficiently with a differential of 3°C to create 1.1 newtons of force.

Strictly, a newton is the amount of force required to accelerate 1 kg at a rate of 1 metre per second per second, but in rough terms, it can be understood as the amount of energy required to lift a kilogram one metre. Clearly, the Beverley clock is not putting out enough energy for useful work, but let it be supposed, that the one cubic foot box were to be 1,000 cubic feet (about the size of a garage) and the temperature differential were to be artificially increased to 20°C, say, by using a concentrating lens, and we are talking about a very usable source of energy indeed, enough to lift a kilo weight to perhaps 30m. This little example takes us to the heart of the K-gen system, for it illustrates the opportunities that exist for producing electricity from renewable sources, once we have a clear idea of the principles underlying the behaviour of energy. While it is true that in a closed system energy can neither be created nor destroyed, as the first law of thermodynamics states, the examples given above illustrate how it can be extracted from natural forces and how the engineer can mimic those forces for the benefit of society.

## 11.5 The K-gen Prototype: A Heuristic Challenge

The schematic of the K-gen system shown in Figures 12 and 16 displays it as a flow chart, with renewable energy as the input and electricity as the output. Ideally, the flow diagram would have quantified the amount of energy produced or lost at each stage of the system, using the same units for all forms of energy, but exact quantification, however desirable, is impossible at this conceptual stage, even with the most sophisticated computer modelling. There are too many variables in the timing and amounts of natural energy input, and the performance of some of the component parts at this stage is uncertain. There is a need to match the capacity of each component, but the critical gap in knowledge is how to match the surplus energy available in the late autumn with the storage capacity of the system in order to meet the demands of the harshest winter. A system which produced optimum usable energy in summer would be completely inadequate in the winter, whereas one scaled to meet winter demands would produce an excess of energy in the summer that would have to be dumped. Seeking an average as a compromise might at first be seen as the answer, but that would still result in a shortfall in winter. As one grapples with a range of uncertainties, it becomes clear that while designing a K-gen system calls for more than usual imagination, building a prototype is a textbook example of heuristic engineering. That is to say, it must rely, at least initially, on approximation, best guess and trial and error, with exceptional margins for error built in at certain places. The first prototype will almost certainly be both over-engineered and under-engineered.

### References

1. Walter Youngquist, "The Post-Petroleum Paradigm," Population and Environment, vol. 20, no.4, March 1999. Also online.

2. Walter S Ong, Orality and Literacy, NY: Routledge, 2002. p. 82.

3. The word was coined by Alcoa (the Aluminum Company of America) and the definition is from an advertisement in Time magazine (24/10/1942) headed, "The Place They Do Imagineering."

4. See, Matthew Dunn, "Scientists claim Anthropocene era started with the New Mexico atomic bomb in in 1945," www.news.com.au.

5. The quote is from an interview with Wilson, entitled "Don't let Earth's tapestry unravel", in *New Scientist*, 24/1/2015.

# Chapter 12

## Raising Awareness

*The human race must finally utilize direct sun power or revert to barbarism.*

Frank Shuman, commercial solar pioneer (1911).[1]

### 12.1 Six Horseman

The book has covered various aspects of energy science, the geopolitics of energy, engineering, and architecture. All of these call for more detailed study than can be given here, and in treating them with brevity it would be easy for the central point to be blurred, that there is a complex global energy crisis and, if the book's thesis is correct, that there is a clean, renewable answer to it, albeit at this stage largely theoretical. The main obstacle to actualizing it lies not in the engineering but in bringing together existing technology and then in motivating individuals and institutions to apply it. The wide sweep of the book's theme may lessen the impact of that message, but the risk has had to be taken because its interwoven themes of climate change, energy crisis and geopolitics are unique in the history of our species and call so urgently for attention. It will be worthwhile, therefore, in this brief concluding chapter to return to some of the themes in summary form and place the energy crisis now looming into a wider context of global problems. This will reinforce the overarching theme that an energy revolution is now called for and is technologically possible.

The complex of crises now facing humankind stem almost entirely from an exponential rise in population, which itself stems from the invention of engines to burn coal and oil and chemical processes to transform oil and gas into fertilizer and pesticides. In the last two centuries we have witnessed unparalleled growth in human well being, but this has resulted in unforeseen developments that now threaten the future of the human race and, quite possibly, of life on planet earth.

Two thousand years ago the writer of the *Book of Revelation* foresaw the end of the world coming and symbolized its causes as four horsemen, representing death, famine, war and its subsequent devastation. This familiar metaphor could now be extended to emphasize that as two centuries of unprecedented growth comes to an end, human civilization faces terminal stress in the following interconnected ways:

- Catastrophic climate change
- Overpopulation
- Exhaustion of many mineral resources
- Permanent food shortage
- Water shortage
- Peak Oil and energy shortage

All these trends are converging towards a point of disaster around the middle of the century, but the energy crisis holds a place of particular importance, for if all the others could be solved, even in imagination, a world without energy such as we have enjoyed unthinkingly for two centuries would be the end of civilization as we have known it. A new kind of energy policy on a global scale is now necessary. Modern civilization has been built on cheap and boundless energy from fossil fuels and until very recently there has been almost no awareness that they are a finite resource and that, thanks to them, we have been living in a completely anomalous period of material riches. Like Cinderella, whose coach was destined to change back into a pumpkin, when the midnight hour of total oil and gas depletion strikes, we shall have to return to a level of material poverty that most in the West have forgotten ever existed – unless we can find a new source of energy on that scale. All this, of course, is assuming tht global warming has not by then wrecked the planet.

The main thrust of the book comes from the argument that such a source is available for the taking if a systematic approach is adopted to harnessing, manipulating and storing energy. A coordinated strategy in both energy engineering and architecture will constitute two revolutions. At the same time we must set about undoing the catastrophic effect on our planet of hydrocarbon sourced energy by extracting and sequestrating atmospheric carbon dioxide. That this can be done by turning it into biochar on an appropriate scale is a third revolution in the making, but calls for such a range of new thinking that it has had to be left out of the main body of the book and its major features outlined in an important appendix. If the argument of this book is accepted, the solution to the global energy problem – in all its aspects - is not primarily one of technology but of imagination, motivation and mobilization of human effort on a scale that few would contemplate, except when it comes to killing other human beings in war.

## 12.2 Thinking Globally

The first chapters of the book were written to raise awareness of the global and national energy crises now building up. In a word, if we continue to use fossil fuel at the rate we have been doing for the past fifty years, carbon dioxide emissions will increase the so-called greenhouse effect until global warming has gone beyond a tipping point. Little was said about the diminishing quality of known reserves of oil and the financial and environmental costs of extracting it. As the extraction of oil from the Canadian tar sands is ramped up, the appalling damage inflicted on hundreds of square kilometres of pristine forests and lakes in northern Alberta is a cost to the planet and the local inhabitants that does not appear on the balance sheet of the giant energy companies that are wreaking such destruction in the cause of profit. However, unless the most drastic action is taken now, long before the world's oil and gas have been exhausted, at whatever environmental cost, we shall have set in train a process of irreversible climate change that will have a devastating impact on agriculture globally, resulting in widespread famine, for as the late professor Albert Bartlett, a long time crusader for sustainability, put it, "Modern agriculture is the use of land to convert petroleum into food." The signs of climatic change are already clear, in more frequent and more destructive hurricanes, droughts, rainfall and flooding on an unprecedented scale. July 2015 was the hottest July globally ever recorded. Unfortunately the general public is only slowly awakening to this unfolding scenario, as some new disaster from flood, wildfire, landslide or tornado makes the news. To make matters worse, there are still individuals, some prominent, who argue that global warming is either not happening or that it is not caused by man-made pollution. Television programmes or newspaper articles will often give equal weight to the case against climate change out of a misplaced sense of balance or fairness. Worse still, some sections of the media, including once respected daily newspapers, periodically mount what appear to be orchestrated character attacks on prominent researchers, using misquotations and distorted data. Is this just a case of "shooting the messenger" because their editors do not want to hear the bad news about climate change, or do not want their readers to hear it, or are there other interests at stake?

That there is uncertainty in the predictions made earlier in the book cannot be denied, for climate change is caused by many feedback loops, short and long term, negative and positive, making precise details and dating impossible. However, as well as the shrinkage of glaciers and Arctic ice albedo, as explained by Professor Wadhams in the foreword, there is a second "smoking gun", namely the correlation between the graphs of rising global temperature and the increase in the use of fossil fuel since the industrial revolution began in earnest two centuries ago. The two curves show an almost perfect fit, both being exponential, with a long period of flatness followed by a gentle rise now turning towards the vertical. Advances in meteorology have also revealed that the jet stream, particularly in the northern hemisphere, is now tending to become stuck in patterns of damaging periods of extreme weather. The endlessly long wet winter of 2013-2014 in the UK and comparable periods of cold and snowstorms in the eastern and central US are two examples of this, and probably the increasing drought conditions in the south western states of America. It needs little imagination to see how food production will be drastically affected by such weather behaviour and is indeed already being affected.

Most alarming of all is the prospect of vast amounts of methane being released from under the shallow waters of the very northerly continental shelves and what is now frozen tundra. Several climate scientists see this threat as so imminent and so great that they have got together on their own initiative to set up the Arctic Methane Emergency Group (AMEG) to serve as an information source, a discussion forum and an action group. The answers so far proposed are on the scale of geo-engineering and are largely variations on seeding the atmosphere with aerosols and reflective particles. The amount of chemicals involved (typically titanium dioxide) would run to million of tons Even if it should be physically possible on such a scale, it is hardly likely to be economically feasible in a world where all governments are now desperately trying to reduce their budgets through "austerity" programmes.

The Sustainable Development Network at the United Nations has drawn up a report of the conclusions of expert teams from fifteen countries, all agreeing that in order to prevent the earth's atmosphere rising above the critical 2°C, as explained in Chapter 2, the annual carbon dioxide emissions per individual would need to be reduced to below 1.6 tonnes. How realistic is this target, considering that it is less than a tenth of American emissions today? Agreement effectively ends there, for after all manner of minor improvements had been factored in as possibilities, from hydrogen fuel cells to electric cars, the most optimistic figure that the group could come up with was 2.3 tonnes. Even that figure was based on what the *New York Times* review called "the most aggressive assumptions,"[2] which is really a euphemism for "wishful thinking". The report's projections of future global temperatures vary widely, from the catastrophic to the unthinkable. The trajectory of climate warming on a best estimate is predicted to be between 2.5°C and 4.8°C, with a worst case of 7.8°C "by the end of the century."[3] Very similar figures have been reached independently by the Chinese State Council's 2011 Second National Assessment Report on Climate Change, its best case being 2.5°C and a most probable of 4.5°C.[4]

It is hardly imaginable what state the planet would be in if the future turned out to be the worst case scenario. In such difficult situations humans tend not to imagine the worst, and throughout the UN document there is a strange wavering between pessimism and optimism. On the one hand, the authors declare that "We do not subscribe to the view held by some that the 2°C limit is impossible to achieve," (p. xiii) but, on the other hand, they concede that "aggregate global models ... are not granular enough to present a detailed technical roadmap for policy implementation at the country level" (p. xv). In other words, strategic solutions are not in evidence. The recommendations of the report, such as they are, put the burden jointly on governments and large businesses but, alas, governments have other priorities, and businesses exist by definition for profit and certainly not to save the world. The proposals of the present book are almost diametrically opposed in relying not on government intervention and subsidy but on householders and small businesses to implement microscale improvements by purchasing as an investment a system that will fulfil their energy needs and give a worthwhile return.

## 12.3 Conclusion: Escaping the Trap

It is difficult to comprehend the size and complexity of the energy problem that now faces the human species and difficult to concentrate attention on it because of the drumbeat of war that grows louder and more confusing. The wars now being waged – including the economic and propaganda wars - and those threatening have obvious ethnic and religious causes but the deepest root of all, as argued in Chapter 3, is the global competition for energy and all the economic and financial consequences that follow.[5] Gaining control of oil has always been the "Great Prize" and the history of the past hundred years has been determined by the fact that most of the world's sweetest and cheapest oil was by geological chance laid down tens of millions of years ago in particular regions of the earth and in some places – like the United States, Mexico and the North Sea – is nearing exhaustion, despite new recovery techniques of fracking and horizontal drilling. Consequently the pressure to gain control over other regional sources increases. A separate study would be needed to tease out the tangled threads of oil politics that have taken the world to its present impasse and while the early chapters of the book have given only a brief overview, it will hopefully suffice to illustrate the impossibility of mobilizing efforts internationally to head off the twin disasters now approaching. Fine words about the "international community" are belied by the realities of nations competing for the hydrocarbons that the earth still holds. Intergovernmental action that has so far been taken, in the form of so-called carbon credits, is worse than a farce: it has turned the energy crisis into a new form of gambling in the financial derivatives casino. The profit-making goes on as the ship of earth steams towards disaster. What this book argues is that the "great prize" is now a renewable energy machine in combination with an architectural revolution in energy-efficient building combined with systematic carbonisation of organic waste.

Without awareness of the problem, no concerted action is possible, and this is why the first action to be taken must be to raise awareness of the inter-related issues. This would be neither a complex nor expensive undertaking for any government, but the difficulty is that politicians speak for their party, not always even in the national interest and, too often, for the corporations that back them financially. (Over 95% of party and campaign funding in the US comes from corporate special interests.) We are left therefore with the question unanswered: who speaks for the global family and the planet? Despite all the obstacles, the first step towards escape from the global energy trap must be to convince the political class that it is they, as our elected representatives, who have responsibility.

What this book offers is a new way of looking at things that will not call for huge capital expenditure. The biggest challenge of all is probably the conversion of existing housing stock to employ the principles that have been outlined in the K-gen system and the E-plus house, but throughout it has been emphasized that the greatest incentive to making an energy revolution happen must be to show the householder how profitable it can be. Although the UK government is operating with a huge annual budget deficit, it could make no more intelligent decision than to subsidize the production of electricity by individual citizens and small private companies through generous grants. As the household's energy costs are reduced, disposable – and thus taxable - income would rise. It would be a win-win situation. The production of biochar through pyrolysis is perhaps not such a clear cut case for grant-aid, but enough will be said in Appendix 1 to show that as well as extracting carbon dioxide rapidly and steadily from the atmosphere its agricultural benefits could be immense, the more so since the world's supply of artificial fertilizers will diminish in line with fossil fuels, eventually almost to nothing.

The world is only now waking up to the threat of global warming and still has not taken on board the social and economic consequences of Peak Oil. The double blow, when oil runs out and atmospheric pollution has passed a tipping point, will almost certainly be upon us in half a century, give or take ten years. It is still understood only by a few and, by and large, the popular press does little to inform the public of the scale of the crisis now developing. If we wait for another fifty years to take action, the damage will have become so frighteningly obvious, that a case for action will not need to be made, but by then it will be too late, as the possibilities for action will have been closed. The message of the book is that the twin problems of energy shortage and global warming are both soluble, but the window for understanding and action is narrowing at a rate that should give us all sleepless nights.

### References

1. Frank Shuman, "Power from Sunshine: A Pioneer Solar Power Plant," *Scientific American*, Sept. 30, 1911.

2. Eduardo Porter, "Blueprints for Taming the Climate Crisis," *New York Times*, July 8, 2014.

3. *Deep Decarbonization Pathway Project.* United Nations Interim 2014 Report. Executive summary, p. xii.

4. Shannon Tiezzi, "In China, Climate Change is Already Here," *The Diplomat*. 14/8/2014.

5. The geopolitical consequences have become harder to predict since the November 27, 2014 meeting of the Organization of Petroleum Exporting Countries (OPEC), when Saudi Arabia used its dominant position as "swing producer" to force a decision not to restrict global oil supply. The reasons for this were not made explicit, but the impact on

various oil-producing countries, notably Russia, Iran and Venezuela, would be very damaging, perhaps catastrophic. What can be said with some certainty is that the oil price is now being used as an instrument of economic war.

# Appendix I

## Biochar and Chemical Sinkage: The Third Strategy

*There is one way we could save ourselves from global warming and that is through the massive burial of charcoal.*

James Lovelock [1]

Of the three strategies that the book proposes for escaping the global energy trap the sequestration of atmospheric carbon dioxide as solid carbon may be the most important. It is, indeed, possible that if a programme were to be implemented with urgency and on a large enough scale, it could resolve the problem of climate change by itself. That it has not been treated in more detail in the body of the book is due in part to the amount of existing material on the topic, much of it undetailed and unrelated, and the social and agricultural consequences that would emerge from a comprehensive policy. This appendix will give only the bare bones of the process which is summarized in the term "biochar", of its feasibility as a geoengineering strategy and its potential effect on farming and industrial practice. More detailed sources of information are given in the end notes.

At the heart of a biochar strategy is the process of pyrolization by which organic and carbon-containing inorganic material is converted to almost pure carbon by heating to a temperature of about 500°C in a low or no oxygen environment, effectively cooked. This has been done for many centuries to make charcoal from wood and after the industrial revolution to make coke from coal, both end products being used as high energy fuels. There is a great deal of technical information on pyrolysis, and retorts or ovens are available commercially in many designs and from household scale to complexes covering half an acre. Biochar can also be produced by microwaving,[2] and although this will probably be the route taken into the future, information here will be restricted to the traditional methods as the best aid to initial understanding of what is at issue.

The particular importance of biochar in the context of the global energy trap stems from four main facts:

- it can remove carbon dioxide from the atmosphere, theoretically on a very large scale
- the charcoal so produced is virtually non-degradable and non-polluting
- it can have very beneficial effects on soil when buried
- its byproducts can go some way towards making good the loss of fuels, plastics, fertilizers and pesticides that will follow from depletion of the earth's oil and gas.

Very large scale sequestration of carbon through pyrolysis breaks into the planetary carbon cycle which the burning of fossil fuels has unbalanced and could in theory rebalance it. The carbon cycle was set up some three billion years ago with the appearance of chlorophyll in biological evolution. Using chlorophyll as a catalyst, green leafed vegetation was able to absorb solar energy and turn plants into organic factories for producing sugars and other more complex chemicals, using the carbon dioxide and hydrogen atoms in water vapour as a feedstock. The carbon and oxygen molecules are separated and rearranged and the oxygen is returned to the atmosphere as a byproduct. This "waste" oxygen is, of course, what keeps humans and all living creatures alive. As vegetable life decays and dies, the carbon dioxide that previously was taken from the atmosphere is returned to it, until eventually it is reabsorbed by growing plants, and the process begins again. Plants also absorb atmospheric carbon dioxide from other sources, such as volcanic events. This continual recycling of carbon has kept the level of atmospheric carbon dioxide at a fairly steady level for many millions of years but with growing emissions from fossil fuel it has risen to the point where life on earth is now threatened.

At present the concentration of carbon dioxide is 389 ppm (parts per million) and the recommendation of the International Panel on Climate change is that it should be reduced as a matter of urgency to 350 ppm and after that as quickly as possible to a pre-industrial figure of about 270 ppm. At this level the carbon cycle is systemically stable, although it could be disturbed temporarily by exceptional volcanic activity. The scale of the challenge now facing the human species may be seen in the Wikipedia entry on Biochar: "The biochar process breaks into the carbon dioxide cycle, as did coal formation hundreds of millions of year ago." What now must be attempted is to lay down in the space of two or three generations carbon deposits on a scale that nature previously did over tens of millions of years. The immediate question is, Can it be done? or, more precisely, Can it be done on the scale required and in the window of time remaining to us? The biggest problem seems to be that no government is prepared to pay producers of biochar for making it or farmers for burying it.

While biochar is already being produced in large quantities and sold commercially,[3] to contemplate extending this to operations on a geoengineering scale raises immediate issues of obtaining the materials for feedstock and disposing of the literally mountains of biochar that would be the end product. Fortunately, the latter does not constitute a problem, since solid carbon is effectively non-degradable, so that any biochar that could not be used productively in agriculture could simply be left in spoil tips and any foreseeable problem from weathering could be solved by covering with a thin overburden of soil.[4]

The soil-improvement effect of biochar is the subject of considerable ongoing research, with some contradictory findings that call for resolution. Benefits seem to depend on the kind of soil, with acidic soil gaining the most. The origin of the feedstock for pyrolysis is also important, for biochar retains some of the structure of the original material which, when made from wood, makes it amenable to colonization by micro-fungal organisms. These occupy the minute spaces of what previously was cellular structure and have a beneficial effect that may broadly be compared to the soil enhancement that worms bring. There is certainly a dramatic increase in crop production when the biochar is "doped" or "charged" with small amounts of normal fertilizers, such as phosphates and nitrates. Not the least agricultural benefit of biochar is that it increases the retention of water in the soil, combating drought and minimizing run off of nutrients and pollution of waterways. One unexpected benefit has been found in using charcoal as a dietary supplement for animals, and particularly cows, where it reduces flatulence and hence the emission of methane, another greenhouse gas.

This short list will be sufficient to indicate that there is great promise in biochar but much still to learn. Leaving aside the critical issues of funding and the daunting logistics of a large scale initiative, the most obvious gap in our knowledge concerns the amount of organic waste that would be required to produce biochar in quantities sufficient to make a worthwhile reduction of atmospheric carbon dioxide. As this question is tackled, it become clear that household, farm and industrial waste will not be sufficient and that it will be necessary to grow crops for the sole purpose of converting to biochar. This raises a whole new set of questions for the agronomist, e.g., what would be the most appropriate crops, what kind of land, what growing time, as well as complex questions about the kind of byproducts that each type of crop would produce along with biochar.

The figures that we have so far give initial hope that we can produce on an ongoing basis the amount of biomass required to rebalance the carbon cycle and, if the challenge were to be attacked with urgency, the very worst effects of global warming could be averted. The excess of atmospheric carbon dioxide is calculated to be in the region of 200 billion tonnes, to which is added each year about eight billion tonnes,[5] and while these figures may at first seem impossibly large, the latter represents only a little more than one tonne of biochar per year for every man, woman and child on the planet, or, very roughly, about 2.8 kilos per person per day, which is capable in theory of being converted to biochar and permanently removed from the carbon cycle. This is somewhat more than the convertible waste that we produce, but at least the figures start to look more manageable. Johannes Lehmann, a pioneer and doyen of the biochar movement has calculated that it is possible to sequestrate 8.5 billion tonnes of carbon per year to offset the current emission from fossil fuels using only available household, agricultural and industrial waste as feedstock.[6] If all this potential were to be actualized, it would reduce the atmospheric load steadily by somewhat less than one per cent per year, which would be nowhere near enough to stabilize, let alone reverse, global warming, and the planet will have burned up long before the excess 200 billion tonnes has been eliminated. Nonetheless, it suggests that more radical thinking about the possibilities could close the gap.

Across the globe there are huge disparities between the amount of waste material generated by households. An Indian peasant will produce almost none and a particularly wasteful western middle class urban family might produce two tonnes per year. The gross figure for the UK is a fairly steady 25 million tonnes. There are about 23 million households in the UK, so each household is creating about one tonne of assorted waste per year, or 20 kilos per week, composed mostly of food and vegetable waste, paper, packaging, glass, metal and assorted plastics. Much of this is already being recycled and from the raw figures it is difficult to get a useful estimate of how much of the total could be converted into biochar. At a guess, perhaps 3 kilos of food and garden waste per week would be treatable and perhaps 6 or 8 kilos of paper and packaging waste could be sequestrated as charcoal but with little or no benefit for the soil. The rest might be unusable plastic residue or metals, some of which could be recycled, but much would probably have to go to landfill. As yet we have no reliable figures.

As well as household waste, a significant but not easily quantifiable amount could come from agriculture by putting two distinct strategies into effect. Since a very large proportion by weight of all crops is inedible, there is potentially a huge amount of vegetable and animal waste that could be diverted into the production of biochar, although much of this is already put to other use. Straw from cereal crops, for instance, is often used as fuel for heaters that dry out wheat grains before milling or composted with animal matter to produce manure. Much of the waste from cow, pig and poultry farms is currently treated as a pollutant rather than as manure but with a planned strategy of pyrolysis could be turned from a liability to a resource. How much of all this could be channelled, even in theory, into the production of biochar is a question that cannot be answered with precise quantities, and probably will never be answerable. We can only start with magnitudes, but these do appear to be promising.

The second possible strategy is growing crops specifically to reduce atmospheric carbon dioxide through pyrolization and sequestration of the biochar. Craig Sams, a former chairman of the Soil Association, has estimated that if (*per impossibile*) all agricultural land in the world were given over to producing biochar, enough carbon dioxide would be taken out of the atmosphere in one year to return it to pre-industrial levels. [7] While that scenario is fantasy, the figures can be extrapolated to show that if more reasonable figures were plugged in – say allocating a fortieth of the land area to biochar production - we could (other things being equal) solve the problem of global warming in forty years. This is a huge simplification of a very complex problem, with several interacting variables but, so far as it goes, gives further cause for optimism.

The main issue of how much land would need to be dedicated to biochar production is dependent on several factors which do not present a well-defined problem, since deciding on the best kind of crop for this purpose calls for the balancing of several factors. Notably among them are the productivity of the chosen crop, how much fertilizer it would need, what byproducts could be obtained and how far the "best" candidate could be grown on marginal land. The biomass required could come from harvesting grasses, shrubs or trees, each of which has quite different characteristics. While these cannot be compared in this short introductory treatment, it should at least be mentioned that shrubs like jatropha and camelina are already used to produce biofuel (see Appendix II) and would, in theory, give a double benefit in a world which will soon run out of petroleum based fuel.

At present there is no central directory where this diverse kind of information could be accessed, although the Biochar Association in the UK, the European Biochar Network and the International Biochar Initiative provide starting points for entry into a fragmented and undeveloped research field. A very wet finger estimate of between 3% and 5% of existing acreage would appear to be sufficient to set aside for biochar production when added to other sources. This figure will doubtless increase as the need for vegetable-based fuels, industrial chemicals and fertilizers increases with the depletion of oil and gas. Traditional charcoal burning never sought to use the so-called volatiles that were driven off in the process, but the pyrolysis of coal in the 19th century opened up new horizons in this field. In the production of coke, syngas is given off and other substances which it was found could be converted into a wide range of other chemicals, such as dyes, bitumens, phenols and even pharmaceuticals. Syngas (called "town gas") was created on a large scale in the 19th and 20th centuries and enabled humans for the first time to have street lighting, then was used for domestic heating and cooking, until replaced in the UK by natural gas from the North Sea. Pyrolysis can be tuned in various ways to provide syngas and varying amounts of valuable chemicals, mostly by regulating the available oxygen to slow or speed up the burn. Typically, slow pyrolysis of 100 kilos of wood waste will produce 30kg of char, 35kg of liquid condensate and 35 kg of gas, as against comparable proportions for fast pyrolysis of 15kg, 55kg and 30kg.

The ramifications of theorizing about a planetary scale biochar programme may be illustrated by a quick look at bamboo as a possible candidate for purpose-grown cropping. Bamboo is actually not a tree but a member of the grass family. There are many quite different subspecies, most being fast-growing, some astonishingly fast, up to a metre a day. Rather than being chopped down and replanted every few years, it can effectively be coppiced, demands little in the way of soil nutrients, but a great deal of water. Its dense root system, allied to the fact that replanting is not required, as with trees after harvesting, slows down the run-off of nutrients and more particularly of phosphates, which pollute waterways with algal blooms. Bamboo yarn is also a good cotton substitute, when the pithy core of the stem is treated much as hemp has been traditionally treated to provide linen yarn. This has to be good news for the earth, for cotton cultivation is one of the most demanding in terms of labour and continual application of fertilizer. Rayon can also be produced from the cellulose of the stem itself, and without doubt other byproducts can be developed, involving chemical processes, some new, some well understood.

As all these factors are considered, it becomes increasingly apparent that what starts as the simple concept of a biochar strategy for sequestrating carbon, ramifies to take in hypothetical issues of many kinds which, as they are tackled, could, or could not, call for a whole new agricultural and industrial strategy. Despite success on a limited scale and the support of almost all prominent individuals in the campaign against global warming, the size and uncertainty of a major biochar initiative, plus the fact that no government or official document appears to support one, leaves many sceptics unconvinced. One of the more surprising is George Monbiot, whose crusading work on global warming the book has cited. As long ago as 2009 he was writing scathing dismissals of biochar as a possible solution to the problem, arguing that the hazards arising from vast plantations outweighed the unproven benefits and ridiculing by name well known supporters like James Hansen and Chris Goodall for being "suckered" by "this latest utopian catastrophe".[8] His case for the opposition is, however, no more firmly based than the arguments of those he criticizes so sarcastically. The plain fact is that without more experimental evidence we simply cannot reach a firm conclusion about the revolutionary potential of biochar. Commonsense would dictate that we press on with obtaining the evidence, for the stakes are too high either just to assume the best unthinkingly or walk away.

In addition to the production of biochar through pyrolysis and microwaving, a method of extracting carbon directly from the air is being researched in several places, most notably at Columbia University in New York, based on bubbling air through water in which an alkaline chemical has been dissolved. Since the carbon dioxide in the air is mildly acidic, it will in theory combine with the dissolved base. In practice, however, there are problems in maximizing the speed and efficiency of the process, specifically in finding the most effective base and catalyst to speed it up. The preferred choice of base so far appears to be sodium hydroxide. Klaus Lackner, the director of the Lenfest Center for Sustainable Energy at Columbia University, has taken this simple idea and scaled it up from the fish tank which was used in his original experiments. One problem, at first unseen, is how to dispose of the end product, or perhaps even market it. The process mimics in some respects the formation by nature of metamorphic rock, and if it were to be developed on the scale needed, there would be huge amounts of "spoil" to be disposed of. Lackner's system, which he calls "engineered chemical sinkage" was ranked by *Discover* magazine as one of seven ideas that could change the world.

One point of particular interest is that if the process is successful on a commercial scale, it will require large amounts of energy and hence a device to provide "free" renewable energy would be a natural partner. Lackner has chosen to use wind energy, a decision which has played a critical part in experimental design. If chemical sinkage were to fulfil its early promise, it could be a partial answer to the problem of dumping surplus electricity from wind farms, as outlined in Chapter 5.

Further information on biochar, in addition to the references below can be found on the websites of the European Biochar Network at www.biochar-europe.org, the International Biochar Initiative at www.biochar-international.org. See also Johannes Lehmann and Stephen Joseph (eds.), *Biochar for Environmental Management: Science and Technology*. London: Routledege, 2009 and Anna Austin, "A New Climate Change Mitigation Tool," *Biomass Magazine*, Oct, 2009. A summary account and assessment of chemical sinkage can be found in M.K. Dubey, et al., "Extraction of Carbon Dioxide from the Atmosphere Through Engineered Chemical Sinkage," Earth and Environmental Sciences Division, MS D462, Los Alamos National Laboratory, BNM87545.

Worth mentioning in the context of biochar, because it is an unnoticed aspect of carbon sequestration, is the long term effect of using wood instead of plastic in many household applications, such as furniture, worktops, shelves. Plastic has clear advantages in many instances, such as window frames, but we should train ourselves to be aware of those situations where wood is an adequate, if slightly more expensive, substitute, particularly perhaps in steel-framed buildings. The net effect on global warming of one household using wood instead of oil-based plastic is obviously negligible, but not if this was adopted as a global strategy.

## References

1. "James Lovelock's One Last Chance to Save Humanity from Climate Change: Burying Large Amounts of Charcoal in the Ground." Interview with Mat McDermott, *New Scientist*, 23/1/2009.

2. Ondrej Masek, et al., "Microwave and slow pyrolysis biochar – Comparison of physical and functional properties. *Journal of Analytical and Applied Physics*. Vol. 100, March 2013, pp. 41-48.

3. Carbon Gold markets it under the brand name "Grochar", which is sold by most garden centres and on Amazon.

4. The stability of charcoal is almost universally accepted, but Mae Wan Ho, the prominent biochemist, is a dissenting voice. In an article "Beware the Biochar Initiative", posted on the ISIS (Institute of Science in Society) website, 7/9/9. She argues forcefully that biochar is not so stable as generally assumed. Her argument needs to be considered. At the very least it cautions against a too simplistic approach.

5. Chris Goodall, *Ten Technologies to Fix Energy and Climate*. London: Green Profile Books. 2008. p.227.

6. The figures, from Johannes Lehmann, are quoted in Alok Jha, "Biochar goes industrial with giant microwaves to lock carbon in charcoal," *The Guardian*, 13/4/2009.

7. James Bruges, *The Biochar Debate* (No. 16 in the Schumacher Briefings), Dartington, Devon: Green Books. p. 57.

8. George Monbiot, "Woodchips with Everything," *Guardian*, 24/3/2009

# Appendix II

## Reducing Pollution from Transport

Although this book puts forward micro- and mesogeneration solutions to the global energy crisis, they would not impact on the pollution arising from transport and many other uses of liquid fuel, as, for instance, tractors in agriculture. There are several important developments in this field that can reduce atmospheric pollution and thus call for mention, as they offer real help in the challenge to slow, halt and in the long run reverse global warming. Each is a major area of technological expertise and research; so what is said here can only be indicative. Nevertheless, they give cause for optimism that the threat of climate change may eventually be contained.

**The Methanol Economy**
About a third of atmospheric pollution comes from burning petrol, diesel, aviation fuel and natural gas. Battery-driven and hybrid cars are already in production, but these are not likely to make a serious difference, partly because the cost factor will be prohibitive, especially when added to the hidden capital expense of providing a distribution and refuelling infrastructure, and partly because the electricity on which they run is likely to originate from a polluting source.

A quite different approach, in the form of the hydrogen economy, has been promoted as a clean and abundant alternative for almost a century and has the immediate advantage that hydrogen can be produced from water by electrolysis using wind-generated electricity.[1] A second great advantage is that the only combustion product from hydrogen is water. There is a great deal of literature on the hydrogen economy, both for and against,[2] and various prototypes of hydrogen-burning vehicles, including working buses and cars and even a Russian Tupolev 154 jet liner converted in 1988 to run one engine on hydrogen. However, despite successful examples like this, there continues to be an obvious lack of large scale implementation. At risk of oversimplifying, the root cause appears to be an insuperable problem of creating an infrastructure of "filling stations". Given the relatively low energy density of hydrogen compared to petrol, over a hundred tankers would be required to transport the same amount of energy for every tanker currently on the road. Furthermore, the structure of the hydrogen atom makes leakage an ever-present problem, and thus compression, containment and transmission difficult. Hydrogen is, however, so attractive an alternative fuel that, rather than abandon the idea altogether, lateral thinking has come up with a workable solution in the form of hydrogen-carrying liquid fuels, most obviously ethanol, methanol and ammonia.

Use of all three is capable of being expanded to provide a complementary strategy to micro- and meso-generation. Ethanol (ethyl alcohol – $C_2H_5OH$) is fast developing as an alternative to, or partial substitute for, petrol. It is widely used in Brazil, where it is produced from sugar cane, the waste biomass being used as fuel to generate heat and power. In the US government subsidies are having a very visible impact on agriculture by tempting farmers to grow corn (maize) to be turned into ethanol. The energy input into growing corn is far greater than the output when used in cars, and while the problem of diminishing oil reserves is very real, using food crops as the feedstock for commercial fuel is a most wasteful and inefficient non-solution. When it comes to controlling climate change, however, the insuperable drawback is that ethanol is a double polluter: not only is carbon dioxide a product of combustion, but the fermentation process by which it is made also produces carbon dioxide.

A more promising solution may be found in growing non-food plants for biofuel, such as jatropha, castor and camelina, the waste from which can be turned into biochar. The former is very much the front runner in this area, capable of growth in relatively marginal soils and of being refined into a biodiesel that is much less polluting than normal diesel fuel. Jatropha oil mixed with normal aviation fuel has been tested with positive results by various airlines, among them Lufthansa and Virgin Airlines, and the International Air Transport Association has recommended a six per cent mixture as a near term objective for all aviation fuel. As the market develops and planting is increased to meet demand, there is good reason to think that jatropha oil will ultimately make a significant contribution to reducing air pollution from the airline industry.

The main rival to ethanol is methanol (methyl alcohol), which is less energy dense (as can be seen from its chemical formula – $CH_3OH$) but has the unique advantage of being able to use carbon dioxide as a feedstock. Thus, although combustion produces carbon dioxide, it is potentially carbon neutral. At present methanol is produced industrially from the fossil hydrocarbon methane, by the well understood Fischer-Tropsch process, but this can be easily modified to use waste carbon dioxide or monoxide along with hydrogen as feedstock. The latter could, ideally, be produced by electrolysis using electricity from clean renewable sources.

The best known advocate of methanol is George Olah, awarded the 1994 Nobel Prize in chemistry, whose book (with associates) *Beyond Oil and Gas: The Methanol Economy* is an introductory *vade mecum* and worth quoting here to illustrate how methanol production mimics nature, and how it can indeed improve upon nature in reducing carbon dioxide pollution by extracting and reusing it, thus adding, as it were, a balancing loop to nature's complex carbon cycle. The problem of global warming arises from the fact that our industrial civilization has created in two centuries a massive unbalancing loop through carbon dioxide emissions. Olah's solution is to copy and improve upon nature's balancing method.

> *Of course Nature, in its process of photosynthesis, captures $CO_2$ by green plants from the air and converts it to water, using the sun's energy and chlorophyll as the catalyst, into new plant life. Thus, plant life replenishes itself by recycling atmospheric $CO_2$. The difficulty is that [this] takes many million years. As we cannot wait that long, we must develop our own chemical recycling processes to achieve it within the required very short time scale.[3]*

Olah points out that since $CO_2$ and hydrogen are the feedstock for this process, a double benefit would come from locating the production sites where there is a regular supply of the former from furnaces, cement factories and other industrial plants, not least from coal-fired power stations. In Olah's words,

> *This would constitute mankind's artificial version of Nature's $CO_2$ recycling via photosynthesis [and] as $CO_2$ is available to everybody on Earth, it would liberate us from the reliance on diminishing and non-renewable fossil fuels and all the geopolitical instability associated with them.* [4]

This whole subject is a fascinating exercise in modern alchemy, in which the elements C, O, N and H are combined and rearranged to create not only fuels with a range of different advantages and drawbacks but plastics too. Individual genius at every level has enabled us to mimic nature and go beyond it to the enormous benefit of human society. It is a province of science and technology which has created engineering at its best, and its achievements are by no means over. [5] The Holy Grail of methanol engineering would be to fit the flues of coal-fired generating stations with devices to capture the polluting exhaust gases and turn the carbon dioxide into methanol. There is no doubt that this is possible, but lack of political will and the uncertain return on capital investment at this point make it unlikely in the foreseeable future.

## The Promise of Ammonia

The third of the hydrogen-based liquid fuels, ammonia, usually comes in for less attention, but a strong case can be made for it, since it is a more effective carrier of hydrogen, its chemical formula $NH_3$ indicating a high hydrogen density and a lack of carbon, which could find its way back into the atmosphere as carbon monoxide or dioxide. Like methanol, it calls for large-scale industrial production along similar lines but using the Haber-Bosch process, one of chemical engineering's most notable achievements. Its greatest advantage from the point of view of climate change is that its only combustion product is water. There is ample information on the "ammonia economy" on the Internet. [6]

The importance of ammonia as a working fluid in the technology of phase change energy has already been touched upon, but it has two other great advantages. In the context of depleting oil and rising population a crisis in food production can be anticipated, arising from a shortage of fertilizer, since artificial nitrate fertilizer, which uses natural gas as a feedstock, also consumes large amounts of fossil fuel in its manufacture. The production of nitrate fertilizer from ammonia, itself produced from atmospheric nitrogen and hydrogen, was the work of Fritz Haber, the controversial genius of physical chemistry, for which he was awarded a Nobel Prize. It has been estimated that two billion people on the planet today exist thanks to his invention, which has enabled agricultural output to be multiplied by several times. Before that mankind had to rely on human and animal manure and the limited deposits of guano (a sort of fossilized bird manure) mostly from South America. As the supply of nitrates diminishes, food production will diminish accordingly, and the world will have to fall back on its previous sources, perhaps including systematic collection of night soil, as it is delicately called. There have been successful experiments, notably in Australia, in collecting and processing human solid waste on a fairly large scale, but a simpler complementary answer may lie in collecting urine and, as it were, reverse engineering the Haber process. The main problem with urine is that the nitrates it contains convert within a day or two to ammonia, and thus it must be used fresh, obviating a large scale strategy for collecting it for direct onward transport to farms. If the nitrate component could be fixed before it turned to ammonia or reconverted after it had done so, a massive new source of fertilizer could be made available. Hopefully some latter day Haber may come along to solve this problem.

The second great value of ammonia is its capability of being used as a clean fuel to replace petrol and diesel. This has been endorsed and encouraged by ASME (the American Society of Mechanical Engineers) and the following information has been taken from their website.

*Ammonia-based fuels offer great potential for universal use. The present disadvantage is that pure ammonia is not suitable for use in high speed engines. Its flame speed is too low. However, ammonia can be doped by environmentally friendly additives, and then be combustible in high speed engines …. Ammoniated fuel will power an engine or burner with very little modification. Thus, the transition to an ammonia-based fuel economy can be as slow or as fast as societal conditions permit.*

## The Commonsense Magic of Turbocharging

Almost from the time of the invention of the petrol and diesel engines turbocharging has been recognized as a way to boost their efficiency. The principle itself could hardly be simpler, entailing no more than rerouting the hot exhaust gas under pressure and using it to drive a compressor which creates forced induction of fuel and oxygen into the cylinder. The first patent for a compressor driven by exhaust gases was given in 1907, and many automotive and aero companies have built engines incorporating it. Its technical and commercial development until fairly recently, however, have been slow, delayed by a complex of difficulties. From an economic perspective, it added to the cost of the car, from an engineering perspective the turbocharger called for development of heat resisting alloys and lubricants, and early models were a fire hazard frowned upon by insurance companies. There was a hidden psychological factor as well, insofar as for many years turbocharging was associated with increasing the performance of racing cars, rather than saving petrol. As always, the cheapness of petroleum worked as a disincentive to research into engineering economics.

Yet the argument in its favour is so obvious that it is hard to see why the turbocharger has remained relatively neglected, for when it is recognized that the waste gases of combustion exit the cylinder at high pressure and at a temperature of 700-800 degrees, it is clear that an inordinate amount of energy, fuel and power is lost in simply letting the exhaust gas vent into the air. It is, in fact, even worse than this, for a great deal of this wasted energy is converted to sound and the silencer which has to be fitted to muffle it creates a back pressure that adversely affects the performance of the engine.

The engineering aspects of turbocharging are of great interest and not difficult to understand. Its performance can be tuned to give extra power at high speeds, which is useful for racing drivers but offers no fuel economy, or to increase bottom end torque, or anything in between, enabling a turbocharger to be fitted which will keep the engine running at its most economical speed without loss of power. Typically, it enables the engine to develop 40% more torque at 1,700 rpm. That figure is hardly less than astonishing, since it translates roughly into a fuel saving of about 25%, compared with a normally aspirated engine of the same capacity, and without loss of performance. The figure is subject to several variables, not least intelligent driving habits, but is probably on the conservative side, for the current Dodge Dart claims that its 1.4 litre turbocharged model generates 160hp, which is exactly the same as its standard 2 litre model, and translates into a 43% saving in fuel. Turbocharging also brings with it other economies, such as smaller engines which save on size, material costs and friction losses.

There has, in fact, been a quiet revolution in the automobile world as the price of petrol has risen and the economic benefits of turbocharging have slowly become recognized. Ongoing research and development has resulted in more compact and reliable superchargers which, combined with direct fuel injection, give quiet and reliable operation as well as improved performance. Several well known car manufacturers now offer turbocharged engines as an option in the UK, and while the statistics worldwide are not easy to uncover, there is an identifiable curve of increase over the past ten years in sales of turbocharged cars and light and heavy trucks. From effectively zero, the overall figures are now, very approximately, 25% in the EU, 12% in the UK and 8% in the US. If that trend continues, as without doubt it will, the future looks decidedly bright. It is surely not being too optimistic to predict that worldwide, as old cars and trucks are replaced by models with newer engines, demand for petroleum for transport could eventually be reduced to two thirds of current demand. That would mark a very substantial reduction in air pollution and a huge step in the campaign against global warming.

The UK government has come in for a good deal of criticism in these pages for its inaction and changes of policy in the face of the national and global energy crisis, so it is only fair to note here that the Department of Transport has created real incentives to cut down on petrol consumption by reducing the road tax on vehicles with smaller engines and carbon dioxide emissions. This seems now to be having a measurable effect in motivating car manufacturers to give more attention to the development and promotion of turbocharged vehicles. The favourable treatment given to owners of small saloon cars means in practice that those with turbocharged engines can find themselves paying as little as £20 a year in road tax or even nothing at all. That is a very worthwhile incentive and seems certain to add momentum to the turbocharger revolution already underway.

### References

1. The biologist J. B. S. Haldane lectured in Cambridge in 1923 on a scheme to produce hydrogen by building "rows of metallic windmills working electric motors …. At suitable distances there will be great power stations where the surplus power will be used for the electrolytic decomposition of water into oxygen and hydrogen. These gases will be liquefied and stored in vast vacuum jacketed reservoirs probably sunk into the ground [and] enable wind energy to be sorted so that it can be expended for industry, transport, heating and lighting as desired." Published as *Daedalus or Science and the Future* (1925) and available on the Internet.

2. cf. Jeremy Rifkin, *The Hydrogen Economy: The Creation of the World-wide Energy Web and the Redistribution of Power on Earth.* Oxford: Blackwell, 2002. Peter Hoffman, *Tomorrow's Energy: Hydrogen Fuel Cells and the Prospects for a Cleaner Planet.* Cambridge, MA: MIT, 2001. Joseph Romm, *The Hype About Hydrogen: Fact and Fiction in the Race to Save the Climate.* NY: Island Press, 2005. Romm was an adviser on energy policy to the Clinton administration.

3. George A. Olah, Alain Goeppert, G. Surya Prakash, *Beyond Oil and Gas: The Methanol Economy.* Weinheim, Germany: Wiley-VCH, 2006. p. 239.

4. Olah, *Op. cit.* p. 244

5. See, e.g., website of the Methanol Institute, www.methanol.org. and the Wikipedia entry, *Methanol Economy.*

6. See, e.g., "Ammonia for Energy Storage", www.PlanMyGreen.com. "Ammonia for High Density Energy Storage," Gottfried Faleschini, et al. "Ammonia-based Solar Energy Storage," Dunn, Burgess & Lovegrove, 3rd International Solar Energy Society Conference, Asia Pacific Region. "Ammonia is the smarter hydrogen," Dana Blankenhorn, www.smartplanet.com . Also, Hal Hodson, "Grab ammonia out of thin air for fuel of the future," *New Scientist*, 6/08/13

# Appendix III

## Thorium: Safe Nuclear for the Future

Large scale development of the K-gen system, the E-plus house and biochar as the answer to the world's need for non-polluting energy is likely to take many years, and in the meantime global warming accelerates. At present a significant part of the world's electricity is supplied by uranium-fuelled nuclear reactors, which do not pollute the atmosphere with carbon dioxide but bring hazards of their own. The disaster at Fukushima, still ongoing, has served to highlight the world's dependency on uranium-fuelled energy. There are 432 nuclear installations in operation worldwide and 65 in various stage of construction (at March 2014) and, despite major advances in safety engineering, the problems of accidental overheating and of safe storage will not go away. Hopes that pebble-bed reactors would overcome these have slowly had to be abandoned. In the words of a 2010 article in *Nature*, "Every nuclear nation in the world has had a programme to commercialize this type of reactor and they all got nowhere."[1]

By contrast, thorium reactors, which operate on quite different principles and shut down automatically in the event of overheating, are gaining plausibility. In principle they simply cannot overheat, as did Chernobyl, and, in addition, the half life of their spent fuel is magnitudes shorter than uranium, thus making them intrinsically safer. Thorium is sometimes referred to as "the magic bullet" in popular literature and the most optimistic claims are made for thorium-based reactors, promising safe, non-polluting and cheap energy for centuries to come.[2] A sober assessment can be found in the TECDOC- 1450 of the International Energy Association Institute (2005), which points out technical difficulties still to be overcome. However, since China and India have both announced that their future energy programmes will be based on thorium reactors, there is good reason to believe that the major ones have been or can be overcome.

Thorium-based generation has been promoted for fifty years and has had eminent supporters, not least the Nobel Laureate Carlo Rubbia, the Director of CERN, who presented a fully worked-out case for it to the European Commission. It is worth asking, therefore, why it has taken so long for its value to be recognized. There are two reasons, both at first surprising, but both illustrating how easily good engineering and the needs of the planet become subordinated to national and commercial interests. The development of thorium reactors in America was effectively abandoned because spent fuel could not be used for military purposes, and depleted uranium was an integral part of America's military programme, as it was of the Soviet Union's. In Europe financing for thorium research was continually blocked because France overruled other members of the European Community in order to protect its vested interest in uranium-generated electricity, where it was a world leader. A brief account of these developments can be found in online news comment by the economist Ambrose Evans-Pritchard,[3] who makes the perceptive observation that unless the West finds government investment to underwrite a thorium-based strategy, we may be seeing "the passing of strategic leadership in energy policy from an inert and status-quo West to a rising technological power [which will] vastly alter the global energy landscape." The article quotes the claim of a respected authority that a presidential decree mandating the rapid building of thorium reactors could put an end to pollution from fossil fuels. However, as this book has emphasized, to assume that the coal and oil industries would passively allow their profits to be threatened by such a policy is wishful thinking. This is one area of the global energy picture that calls urgently for public debate and enlightenment. An up to date and wide-ranging contribution to the debate can be found in Richard Martin's *Superfuel: The Green Energy Source for the Future*.[4]

.

## References

1. Linda Nording, "Pebble-bed reactor gets pulled," *Nature* 463, 1008-9 (2010.

2. See, e.g., Helen Brown, "Is thorium the answer to our energy crisis," *The Independent.* 13/12/2006

3. Ambrose Evans-Pritchard, "Obama could kill fossil fuels overnight with a nuclear dash for thorium," *The Telegraph*, 29/8/2010.

4. Richard Martin, Superfuel: *The Green Energy Source for the Future.* London: Palgrave Macmillan, 2012

# Index

## A

ACESA 47
Aerogel 95, 163, 187, 188, 189
Agriculture 36
Ahmed, Nafeez 61
Ainsworth, Peter 66
Aircrete 189
Albedo xii, xiv, xviii
Al Qaeeda 52
Ammonia 37, 78, 238, 240, 241, 242, 245
Anderson, Kevin 43
Antarctic ice 3
Anthropocene Era 210, 211, 215
Anthropocene Working Group 211
Arctic ice xii, xiii
Arctic Methane 221
Armstrong, William 22
Arnold, Matthew 59
Arrhenius 108
ASME (the American Society of Mechanical Engineers) 241
Atmos clock 213
Atrium Principle 182

## B

Bamboo 233
Bangladesh 3
Banham, Reyner 173, 177, 201, 202
Barker, Greg 116
Bartlett, Albert 49, 219

Batteries 152
Bauhaus 169, 171
Beckett, Margaret 66
Bernoulli's theorem 212
Betz law 92, 130
Beverley clock 214
Bhadrakumar, M. K. 60
Bingham Canyon 18
Biochar 2, 227, 229, 230, 231, 233, 234
Biochar Association 232
Birkett, Derek 68
Blair, Tony 64
Booker, Christopher 79
Boulton, Matthew 13, 34
BP 54
British Wind Energy Association (BWEA) 74
Brooks, Michael 30
Bruges, James 236
Brunel University 165
Brzezinski, Zbigniew 59, 60
Buchanan, Alistair 64
Buffett, Warren 47
Building Regulations 149, 188

## C

CAD 170
California Wind Systems 134
Camelina 232, 238
Canadian tar sands 219
Cantarell field 48
Carbon cycle 228, 230

Carnot, Sadi 107
casa giratoria 169
Castor 238
Central Electricity Generating
    Board (CEGB) 65
Central heating 195
Chadwick, Edwin 168
Chapelon 161
Chardin, Teilhard de 210
Chemical sinkage 235
Cheney, Dick 52
China 37, 80
Christoff, Peter 44, 50
Churchill, Winston 51, 58, 59
Clark, Wesley 53, 59
Clean Air Act 6, 168
Clean Air and Security Act, The
    47
Climate change 217, 218
Climate warming 221
Coanda effect 130
Control system 96
Conway, Erik M 50
Cox, Peter 49
Crutzen, Paul 210

**D**

Darmstadt 25
Darrieus 132
Dehumidifiers 200
Denmark 72
Denzer, Anthony 201
Department of Energy 119
Department of Energy and Cli-
    mate Change 117
Depleted uranium 248
Desertec 114, 123
Desertec Initiative 27

Douglas, Ed 80

**E**

Eddington, Sir Arthur 28, 29
Edison, Thomas 127
Electrochemical storage 152
Emergency Group (AMEG) 221
Energy 16
Energy Cache 77, 80
Energy Density 19
Energy equivalence 21
Energy, kinetic 21
Energy loss 26
Energy Management 22, 103, 107
Energy, nature of 14
Energy Saving Trust 123, 157
Energy storage 150
Energy Trust 95
Engdahl, F. William 59
Entropic 103
Entropy 14
Environmental Change Institute's
    (ECI) 75
Ethanol 36, 238
Etherington, John 79
European Biochar Network 232
Evans-Pritchard, Ambrose 248,
    249
Exxon Mobil 50

**F**

Feed-in Tariff (FIT) 111, 112, 119
Fertilizers 224
Fischer-Tropsch process 239
Fisher, Richard 27
Flexicity 98
Flybrid 163
Flywheel 78, 95, 153, 154, 155,

161, 163, 192
Flywheel storage 155
Foster, Baron Norman 1, 9
Fracking 5
Fraunhofer and Dardesheim projects 25
Fraunhofer Institute 176
Fresnel lens 142
Fuel cell 95
Fukushima 247

G

Gates, Bill 77
Gemasolar plant 113
Gemini house 169
Geoengineering xviii
Geopolitics 51
Geothermal power 24
Germany 77
Ghawar 48
Giedeon, Sigfried 171, 181, 202
Global warming 36, 227
Goeppert, Alain 245
Goodall, Chris 234, 236
Gore, Al 2
Gossop, John 49
Gravity 106
Great Barrier Reef 44
Great Game 56
Great prize, the 55
Greenland ice sheet xiv
Greenpeace UK 98, 100
Green Sahara 45
Gresley 161
Grid 88, 96, 99, 150, 158
Grid design 26
Grids, very smart 98
Griffin, David Ray 49

Gropius, Walter 169, 171, 201
Gulf Stream 4

H

Haber-Bosch process 240
Haber, Fritz 241
Halbach array 153, 163
Haldane, J. B. S. 244
Halkema, J. A. 79
Hansen, James 41, 49, 234
Harvesting water 196
Hastings, Max 68
Heating, ventilation, air-conditioning system (HVAC) 156
Heat pipes 139
Heat pump 93, 107, 120, 135, 138, 139, 193, 194, 208
Heinberg, Richard 49
Heliostat 141
Hewlett Packard 87
Hoffman, Peter 244
Holocene Era 210
Homeostasis 105
Horns Rev field 71
Hubbert, Marion King 7, 13, 28, 29
Huhne, Chris 73
Humphrey, Colin 185
Hyde, William 79, 80
Hydroelectric 23
Hydrogen economy 29, 237

I

Imagineering 209, 215
Induction motor 94
Industrial Revolution 38
Iniewski, Krzysztof (Kris) 100
Institution of Civil Engineers 64

Insulation 149
Intergovernmental Panel on Climate Change 38
International Air Transport Association 238
International Biochar Initiative 232
International Energy Agency 63
International Panel on Climate Change (IPCC) 3
IPCC 3, 38
Isentropic 151, 163
ISIS 58, 61

**J**

Jatropha 232, 238
Jet stream xvi
Journal of Solar Energy Engineering 141
JP Morgan bank 47
Judicial Watch 58

**K**

Kaiser Wilhelm II 54
Kalgoorlie Super Pit, The 18
Keck, Fred 169, 176
Kinetic energy 21
King, David 31, 41
Kirk, Geoffrey 9
Koch Industries 50
Kunstler, James Howard 31
Kyoto Protocol 47

**L**

Lackner, Klaus 235
La Rance barrage 24
Larder 193
Latent energy 137

Lateral thinking 196
Lawson, Lord 50
LED Lighting 185, 186, 202
Leger, Eugene 172
Leggett, Jeremy 6, 9, 118
Lehmann, Johannes 230, 235, 236
Lenfest Center for Sustainable Energy 235
Libeskind, Daniel 170
Lithium 163
Lithium-ion battery 153
Lloyd Wright, Frank 176
London Shard 171
Lovelock, James 69, 227, 235
Lucas, Caroline 44

**M**

MacKay, David 21, 49
Mackinder, Halford 56
Macrogeneration 24, 113
Mae Wan Ho 236
Mark Serreze xii
Martin, Richard 248, 249
McCarthy, Natasha 85, 101, 167
McKibben, Bill 41
Meacher, Michael 51
Media Research Centre 50
Merchants of Doubt 29
Merz, Charles 22
Mesogeneration 37, 88, 154
Methane xii, xiii, xv
Methanol 37, 238, 239, 240
Methanol Economy 237
Methanol Institute 245
Microgeneration 22, 37, 140
Microwave 236
Milliband, Ed 66

Modermott, Nick 79
Mojave Desert 113
Monbiot, George 13, 15, 16, 28, 29, 98, 100, 234, 236
Morgan, Emslie 177, 178
Mouchot, Augustin 207, 208
Murray-Darling river system 44

**N**

NASA 27
National Grid 6, 22, 25, 63, 65, 67, 73, 74, 93, 97, 111, 112
NATO 52, 59, 60
Neodymium 76, 94
Newcomen 211
Newcomen, Thomas 13
North Atlantic Treaty Organization 56
North Sea field 32
Northwest Passage xi

**O**

Oceanic warming 38
Off-grid homes 140
Ofgem 64, 66, 71, 73, 105, 117
Ofgem report 63
Olah, George 239, 245
Ong, Walter 207, 211, 215
OPEC 177, 224
Oreskes, Naomi 50
Ottoman empire 54

**P**

Page, Michael le 9
Parasitic loss 103
Parry, Simon 80
Parsons, Charles 22

Passivhaus 91, 145, 147, 148, 151, 156, 172, 173, 174, 195
Passivhaus Institut 174
Passivhaus Retrofit 203
Peak Oil 31, 218, 224
Pebble-bed reactors 247, 248
Pepe Escobar 60
Permafrost 40
Philosophy of engineering 101
Photovoltaic 91
Photovoltaic panels 127, 179
Piano, Renzo 171
Pine beetle 39
Prakash, G. Surya 245
Project Discovery 66
Pyrolysis 227, 231, 233

**R**

Radiators 193, 194
Rainwater Harvesting Association 199
Rainwater harvesting systems (RWH) 197
Rankine cycle, organic 142
Rankine engine 180
Realpolitik 52
Renewable Energy Foundation 73
Renewable Energy UK (REUK) 132
Renewable Heat Incentive 118
Renewable Heat Initiative (RHI) 117, 120
Renewable Obligation Certificates (ROCS) 75
Renewable Obligation legislation 76
Retrofit 195, 196

Retrofitted system 122
Retrofitting 86
Return to Equilibrium (RTE),
    principle of 107
Ridgeblade 134
Rifkin, Jeremy 25, 29, 51, 244
Ripple control 158
Riverine power 24
Romm, Joseph 244
Rosenbloom, Eric 80
RTE principle 139, 212, 213
Rubbia, Carlo 248
Rughani, Deepak 49
RWH Systems 197, 198

**S**

Saddam Hussein 57
Sagan, Dorion 109
Sampson, Geoffrey 59, 60
Saskatchewan Conservation House
    172
Saudi Arabia 55, 224
Savery, Thomas 211
Savonius model 132
Scheer, Hermann 59, 60, 115,
    123
Schneider, Eric 109
Schumacher, Ernst (Fritz) 1, 6,
    7, 8
Self-insulated panel (SIP) 188
Semi-basement 192
Sequestrating carbon 234
Shanghai Cooperation Organiza-
    tion 56
Shuman, Frank 207, 217, 224
Siemens 70, 74, 79, 114
Sihwa Lake 24
Silk Road, the new 57

Simmons, Matthew R. 48
Simpson, Alan 74
Sinden, Dr Graham 75
Sloan, Howard 176
Smart Grid 98
Smart Grid Technology 27
Soane, Sir John 181
Socrates 175
Soddy, Frederick 11
Solar dish 141
Solar panel 91
Solar storms 27, 28
Solar thermal panels 120, 128,
    179
Spence, Sir James 168
Stanier 161
Stanton, William 49
State change 137, 138
Stern, Nicholas 42
Stern Report 38
Stevenson, Struan 73
Stirling engine 140, 142, 213
Systems thinking 102
Szent-Gyorgyi, Albert 83

**T**

TAPI pipeline 57
Tarmac 35
Telkes, Maria 169, 176
Tesla, Nikola 83, 107, 125
Thanet wind farm 72
Thermal efficiency 147
Thermohaline circulation xvi
Thorium 20, 112, 247, 248, 249
Three Mile Island 177
tidal power 24
Transnational companies 46
Triple glazing 147, 185

Trombe, Félix  169, 176
Turbocharging  242, 243
TWECS  133, 134

**U**

Underfloor heating  193, 194
Uninterruptible Power Supply
    systems (UPS)  99, 155
University of East Anglia Climatic
    Research Unit  45

**V**

Vacuum tube  91
Vattenfall  73
Venus syndrome  41
Vernadsky, Vladimir  210
Voltage Optimization  156

**W**

Wadhams, Peter  49
Warwick Wind Trials  129, 135
Wasdell, David  49
Waste heat recovery  93, 190
Waste heat retrieval  156
Waste water  199
Water Aid  200
Water consumption  197
Water Supply Regulations  198
Watt, James  13
Watts  211
Webley, John  79, 80
Whyalla  137
Wilson, E. O.  212
Wind energy  94, 129
Wind farms  77
Window reveals  184
Windpods  134
Wind turbines  131, 161

World Health Organization  200
World Meteorological Organisa-
    tion  39, 44
Wright brothers  161

**Y**

Yen, James  134
Yergin, Daniel  59, 60
Youngquist, Walter  205, 215

**Z**

Zero Energy Windows  162
Zone heating  196

www.ingramcontent.com/pod-product-compliance
Lightning Source LLC
Chambersburg PA
CBHW071549210326
41597CB00019B/3171